NANOTECHNOLOGY SCIENCE AND TECHNOLOGY

CARBON NANOTUBES AND NANOFIBERS

OCCUPATIONAL EXPOSURE RISKS AND MINIMIZATION STRATEGIES

MIRIAM A. SUMMERFIELD
EDITOR

nova publishers
New York

For permission to use material from this book please contact us:
Telephone 631-231-7269; Fax 631-231-8175
Web Site: http://www.novapublishers.com

Library of Congress Cataloging-in-Publication Data

ISBN: 978-1-63117-629-6

Published by Nova Science Publishers, Inc. † New York

NANOTECHNOLOGY SCIENCE AND TECHNOLOGY

CARBON NANOTUBES AND NANOFIBERS

OCCUPATIONAL EXPOSURE RISKS AND MINIMIZATION STRATEGIES

NANOTECHNOLOGY SCIENCE AND TECHNOLOGY

Additional books in this series can be found on Nova's website under the Series tab.

Additional e-books in this series can be found on Nova's website under the e-book tab.

CONTENTS

PREFACE

Results from recent animal studies indicate that carbon nanotubes (CNT) and carbon nanofibers (CNF) may pose a respiratory hazard. CNTs and CNFs are tiny, cylindrical, large aspect ratio, manufactured forms of carbon. There is no single type of carbon nanotube or nanofiber; one type can differ from another in shape, size, chemical composition (from residual metal catalysts or functionalization of the CNT and CNF) and other physical and chemical characteristics. Such variations in composition and size have added to the complexity of understanding their hazard potential. Occupational exposure to CNTs and CNFs can occur not only in the process of manufacturing them, but also at the point of incorporating these materials into other products and applications. This book reviews the animal and other toxicological data relevant to assessing the potential non-malignant adverse respiratory effects of CNT and CNF; provides a quantitative risk assessment based on animal dose-response data; proposes a recommended exposure limit (REL) of 1 μg/m3 elemental carbon as a respirable mass 8-hour time-weighted average (TWA) concentration; and describes strategies for controlling workplace exposures and implementing a medical surveillance program.

Chapter 1 - The Occupational Safety and Health Act of 1970 (Public Law 91-596) was passed to assure safe and healthful working conditions for every working person and to preserve their human resources. This Act charges the National Institute for Occupational Safety and Health (NIOSH) with recommending occupational safety and health standards and describing exposures that are safe for various periods of employment, including (but not limited to) the exposures at which no worker will suffer diminished health, functional capacity, or life expectancy because of his or her work experience.

NIOSH issues Current Intelligence Bulletins (CIBs) to disseminate new scientific information about occupational hazards. A CIB may draw attention to a formerly unrecognized hazard, report new data on a known hazard, or disseminate information about hazard control. CIBs are distributed to representatives of academia, industry, organized labor, public health agencies, and public interest groups, as well as to federal agencies responsible for ensuring the safety and health of workers.

NIOSH is the leading federal agency conducting research and providing guidance on the occupational safety and health implications and applications of nanotechnology. As nanotechnology continues to expand into every industrial sector, workers will be at an increased risk of exposure to new nanomaterials. Today, nanomaterials are found in hundreds of products, ranging from cosmetics, to clothing, to industrial and biomedical applications.

These nanoscale-based products are typically called "first generation" products of nanotechnology. Many of these nanoscale-based products are composed of engineered nanoparticles, such as metal oxides, nanotubes, nanowires, quantum dots, and carbon fullerenes (buckyballs), among others. Early scientific studies have indicated that some of these nanoscale particles may pose a greater health risk than the larger bulk form of these materials.

Results from recent animal studies indicate that carbon nanotubes (CNT) and carbon nanofibers (CNF) may pose a respiratory hazard. CNTs and CNFs are tiny, cylindrical, large aspect ratio, manufactured forms of carbon. There is no single type of carbon nanotube or nanofiber; one type can differ from another in shape, size, chemical composition (from residual metal catalysts or functionalization of the CNT and CNF) and other physical and chemical characteristics. Such variations in composition and size have added to the complexity of understanding their hazard potential. Occupational exposure to CNTs and CNFs can occur not only in the process of manufacturing them, but also at the point of incorporating these materials into other products and applications. A number of research studies with rodents have shown adverse lung effects at relatively low-mass doses of CNT and CNF, including pulmonary inflammation and rapidly developing, persistent fibrosis. Although it is not known whether similar adverse health effects occur in humans after exposure to CNT and CNF, the results from animal research studies indicate the need to minimize worker exposure.

Chapter 2 - Since 1987, the Worker Education and Training Program has been a major national force for developing effective occupational safety and health training. The program has generated guidance on training hazardous waste workers under OSHA's 1910.120 Hazardous Waste Operations and Emergency Response (HAZWOPER) Standard, on using advanced training technologies to train workers, and on training workers exposed to mold. These guidance documents advocate peer training of workers and good use of adult learning techniques.

In the Fall of 2009, the program held a technical meeting entitled, *Global Safety and Health Issues and their Impact on Worker Training*. A panel that included Drs. Kristen Kulinowski from Rice University, Richard Niemeier from the National Institute for Occupational Safety and Health, and Sam Paik from Lawrence Livermore National Lab addressed nanotechnology and control banding and the implication for worker training. The panel was moderated by Dr. Bruce Lippy.

This paper is a direct result of the conference. Rather than produce their customary conference report, the authors felt the subject of training workers about the risks and benefits of nanomaterials was so important and unexamined that the authors should explore it in-depth in a discussion paper that can hopefully lead to a guidance document with broad value to workers and employers. To that end, the authors welcome your comments. Unfortunately, the history of introducing new materials like asbestos, lead, acrylonitrile and polychlorinated biphenyls has been a legacy of government actions *after* disease has been discovered in a worker population. The U.S. government has learned from history and has been much more proactive in identifying the hazards of nanomaterials. The efforts to collect and disseminate nano-related information through outlets like Rice University's International Council on Nanotechnology, the National Nanotechnology Initiative of the federal government that coordinates the activities of 25 Federal agencies, and the Woodrow Wilson International Center for Scholars, has been unprecedented.

The growth of nanotechnologies is so explosive, however, that regulatory agencies are having difficulties keeping up. As the Government Accountability Office (GAO) starkly noted in its May 2010 report on nanotechnology, "Although the body of research related to nanomaterials is growing, the current understanding of the risks posed by these materials is limited." The area that has received the least amount of attention is the training of the nanotechnology workforce. Hopefully, this document will begin to redress that deficit.

In: Carbon Nanotubes and Nanofibers
Editor: Miriam A. Summerfield

ISBN: 978-1-63117-629-6

Chapter 1

OCCUPATIONAL EXPOSURE TO CARBON NANOTUBES AND NANOFIBERS*

National Institute for Occupational Safety and Health

FOREWORD

The Occupational Safety and Health Act of 1970 (Public Law 91-596) was passed to assure safe and healthful working conditions for every working person and to preserve our human resources. This Act charges the National Institute for Occupational Safety and Health (NIOSH) with recommending occupational safety and health standards and describing exposures that are safe for various periods of employment, including (but not limited to) the exposures at which no worker will suffer diminished health, functional capacity, or life expectancy because of his or her work experience.

NIOSH issues Current Intelligence Bulletins (CIBs) to disseminate new scientific information about occupational hazards. A CIB may draw attention to a formerly unrecognized hazard, report new data on a known hazard, or disseminate information about hazard control. CIBs are distributed to representatives of academia, industry, organized labor, public health agencies, and public interest groups, as well as to federal agencies responsible for ensuring the safety and health of workers.

NIOSH is the leading federal agency conducting research and providing guidance on the occupational safety and health implications and applications of nanotechnology. As nanotechnology continues to expand into every industrial sector, workers will be at an increased risk of exposure to new nanomaterials. Today, nanomaterials are found in hundreds of products, ranging from cosmetics, to clothing, to industrial and biomedical applications. These nanoscale-based products are typically called "first generation" products of nanotechnology.

Many of these nanoscale-based products are composed of engineered nanoparticles, such as metal oxides, nanotubes, nanowires, quantum dots, and carbon fullerenes (buckyballs),

* This is an edited, reformatted and augmented version of Current Intelligence Bulletin 65, DHHS (NIOSH) Publication No. 2013-145, dated April 2013.

among others. Early scientific studies have indicated that some of these nanoscale particles may pose a greater health risk than the larger bulk form of these materials.

Results from recent animal studies indicate that carbon nanotubes (CNT) and carbon nanofibers (CNF) may pose a respiratory hazard. CNTs and CNFs are tiny, cylindrical, large aspect ratio, manufactured forms of carbon. There is no single type of carbon nanotube or nanofiber; one type can differ from another in shape, size, chemical composition (from residual metal catalysts or functionalization of the CNT and CNF) and other physical and chemical characteristics.

Such variations in composition and size have added to the complexity of understanding their hazard potential. Occupational exposure to CNTs and CNFs can occur not only in the process of manufacturing them, but also at the point of incorporating these materials into other products and applications. A number of research studies with rodents have shown adverse lung effects at relatively low-mass doses of CNT and CNF, including pulmonary inflammation and rapidly developing, persistent fibrosis. Although it is not known whether similar adverse health effects occur in humans after exposure to CNT and CNF, the results from animal research studies indicate the need to minimize worker exposure.

This NIOSH CIB, (1) reviews the animal and other toxicological data relevant to assessing the potential non-malignant adverse respiratory effects of CNT and CNF, (2) provides quantitative risk assessment based on animal dose-response data, (3) proposes a recommended exposure limit (REL) of 1 $\mu g/m^3$ elemental carbon as a respirable mass 8-hour time-weighted average (TWA) concentration, and (4) describes strategies for controlling workplace exposures and implementing a medical surveillance program. The NIOSH REL is expected to reduce the risk for pulmonary inflammation and fibrosis. However, because of some residual risk at the REL and uncertainty concerning chronic health effects, including whether some types of CNTs may be carcinogenic, continued efforts should be made to reduce exposures as much as possible.

Just prior to the release of this CIB NIOSH reported at the annual meeting of the Society of Toxicology [03/11/2013] preliminary findings from a new laboratory study in which mice were exposed by inhalation to multi-walled carbon nanotubes (MWCNT) [see http://blogs.cdc.gov/niosh-science-blog/2013/03/mwcnt/]. The study was designed to investigate whether MWCNT have the potential to initiate or promote cancer. Mice receiving both an initiator chemical plus inhalation exposure to MWCNT were significantly more likely to develop tumors (90% incidence) and have more tumors than mice receiving the initiator chemical alone. These results indicate that MWCNT can increase the risk of cancer in mice exposed to a known carcinogen. The study did not indicate that MWCNTs alone cause cancer in mice.

This research is an important step in our understanding of the hazards associated with MWCNT, but before we can determine whether MWCNT pose an occupational cancer risk, we need more information about workplace exposures, the types and nature of MWCNT being used in the workplace, and how that compares to the material used in this study. Research is underway at NIOSH to learn more about worker exposures and the potential occupational health risks associated with exposure to MWCNT and other types of CNTs and CNFs. As results from ongoing research become available, NIOSH will reassess its recommendations for CNT and CNF and make appropriate revisions as needed.

NIOSH urges employers to share this information with workers and customers. NIOSH also requests that professional and trade associations and labor organizations inform their members about the potential hazards of CNT and CNF.

John Howard, M.D.
Director, National Institute for Occupational
Safety and Health
Centers for Disease Control and Prevention

EXECUTIVE SUMMARY

Overview

Carbon nanotubes (CNTs) and nanofibers (CNFs) are some of the most promising materials to result from nanotechnology. The introduction of these materials and products using them into commerce has increased greatly in the last decade [Thostenson et al., 2001; Invernizzi 2011]. The development of CNT-based applications in a wide range of products is expected to provide great societal benefit and it is important that they be developed responsibly to achieve that benefit [Sanchez et al., 2009; Schulte et al 2012]. Worker safety and health is a cornerstone of responsible development of an emergent technology because workers are the first people in society to be exposed to the products of the technology and the workplace is the first opportunity to develop and implement responsible practices.

In this Current Intelligence Bulletin, NIOSH continues its long-standing history of using the best available scientific information to assess potential hazards and risks and to provide guidance for protecting workers. Since it is early in the development of these materials and their applications, there is limited information on which to make protective recommendations. To date, NIOSH is not aware of any reports of adverse health effects in workers using or producing CNT or CNF. However, there are studies of animals exposed to CNT and CNF that are informative in predicting potential human health effects consistent with ways in which scientists traditionally have used such data in recommending risk management strategies. NIOSH systematically reviewed 54 laboratory animal studies, many of which indicated that CNT/CNF could cause adverse pulmonary effects including inflammation (44/54), granulomas (27/54), and pulmonary fibrosis (25/54) (Tables 3–1 through 3–8). NIOSH considers these animal study findings to be relevant to human health risks because similar lung effects have been observed in workers exposed to respirable particulates of other materials in dusty jobs [Rom and Markowitz 2006; Hubbs et al., 2011]. There are well established correlations between results of animal studies and adverse effects in workers exposed to particulates and other air contaminants [NIOSH 2002, 2006, 2011a, b]. Moreover, in animal studies where CNTs were compared with other known fibrogenic materials (e.g., silica, asbestos, ultrafine carbon black), the CNTs were of similar or greater potency [Lam et al., 2004; Muller et al., 2005; Shvedova et al., 2005; Murray et al., 2012], and the effects, including fibrosis, developed soon after exposure and persisted [Shvedova et al., 2005, 2008; Porter et al., 2010; Mercer et al., 2011]. These are significant findings that warrant protective action. NIOSH conducted a quantitative assessment of risk using the animal studies with sufficient dose-response data, which included two subchronic (90-day) inhalation studies [Ma-Hock et al., 2009; Pauluhn 2010a] and five additional studies [Lam et al., 2004; Muller et al., 2005; Shvedova et al., 2005,2008; Mercer et al., 2011] conducted by other routes or durations. The estimated risk of developing earlystage (slight or mild) lung effects over a working lifetime if exposed to CNT at the analytical limit of quantification (NIOSH Method 5040) of 1 μg/m^3 (8-hr time-weighted average [TWA] as respirable elemental carbon) is approximately 0.5% to 16% (upper confidence limit estimates) (Table A–8). In addition, the working lifetime equivalent estimates of the animal no observed adverse effect level (NOAEL) of CNT or CNF were also near 1 μg/m^3 (8-hr TWA) (Sections A.6.3.3 and A.7.6). Therefore, NIOSH recommends that exposures to CNT and CNF be kept below the recommended exposure limit (REL) of 1 μg/m^3 of respirable elemental carbon as an 8-hr TWA. Because there may be other sources of elemental carbon in the workplace that could interfere in the determination of CNT and CNF exposures, other analytical techniques such as transmission electron microscopy are described that could assist in characterizing exposures. Studies have shown that airborne background (environmental and in non-process areas in the workplace) concentrations to elemental carbon are typically less than 1 μg/m^3 and that an elevated exposure to elemental carbon in the workplace is a reasonable indicator of CNT or CNF exposure [Evans et al., 2010; Birch 2011a, b; Dahm et al., 2011]. Studies have also shown in some manufacturing operations that exposures can be controlled below the REL when engineering controls are used [Dahm et al., 2011]. However, NIOSH has not assessed the extent to which exposures can be controlled during the life cycle of CNT/CNF product use, but since airborne CNT/CNF behave as classical aerosols, the control of worker exposures appears feasible with standard exposure control techniques (e.g., source enclosure, local-exhaust ventilation) [NIOSH 2009a]. Previously in a 2010 draft of this CIB for public comment, NIOSH indicated that the risks could occur with exposures less than 1 μg/m^3 but that the analytic limit of quantification was 7 μg/m^3. Based on subsequent improvements in sampling and analytic methods, NIOSH is now recommending an exposure limit at the current analytical limit of quantification of 1 μg/m^3.

More research is needed to fully characterize the health risks of CNT/CNF. Longterm animal studies and epidemiologic studies in workers would be especially informative. However, the toxicity seen in the shortterm animal studies indicates that protective action is warranted. The recommended exposure limit is in units of mass/unit volume of air, which is how the exposures in the animal studies were quantified and it is the exposure metric that generally is used in the practice of industrial hygiene. In the future, as more data are obtained, a recommended exposure limit might be based on a different exposure metric better correlated with toxicological effects, such as CNT/CNF number concentration [Schulte et al., 2012].

There are many uncertainties in assessing risks to workers exposed to CNT/CNF. These uncertainties, as described and evaluated in this document, do not lessen the concern or diminish the recommendations. Other investigators and organizations have been concerned about the same effects and have recommended occupational exposure limits (OELs) for CNT within the range of 1–50 μg/m^3 [Nanocyl 2009; Aschberger et al., 2010; Pauluhn 2010b; Nakanishi (ed) 2011a,b]. The relative consistency in these proposed OELs demonstrates the need to manage CNT/CNF as a new and more active form of carbon. To put this in perspective, since there is no Occupational Safety and Health Administration (OSHA) permissible exposure limit (PEL) for CNT/CNF, the PEL for graphite (5,000 μg/m^3) or carbon black (3,500 μg/m^3) [NIOSH 2007] might inappropriately be applied as a guide to control worker exposures to CNT/CNF. Based on the information presented in this document, the PELs for graphite or carbon black would not protect workers exposed to CNT/CNF.

The analysis conducted by NIOSH was focused on the types of CNT and CNF included in published research studies. Pulmonary responses were qualitatively similar across the various types of CNT and CNF, purified or unpurified with various metal content, and different dimensions [Lam et al., 2004; Shvedova et al., 2005, 2008; Muller et al., 2005; Ma-Hock et al., 2009; Pauluhn 2010a; Porter et al., 2010; Mercer et al., 2011; Murray et al., 2012; DeLorme et al., 2012]. The fibrotic lung effects in the animal studies developed early (within a few weeks) after exposure to CNT or CNF, at relatively low-mass lung doses, and persisted or progressed during the post-exposure followup (~1–6 months) [Shvedova et al., 2005, 2008; Mercer et al., 2008; Porter et al., 2010; Pauluhn 2010a; Murray et al., 2012]. However, the studied CNT and CNF only represent a fraction of the types of CNT and CNF that are, or will be, in commerce and it is anticipated that materials with different physical and chemical parameters could have different toxicities. At this time, however, given the findings in the published literature, NIOSH recommends that exposures to all CNT and CNF be controlled to less than 1 μg/m^3 of respirable elemental carbon as an 8-hr TWA, and that the risk management guidance described in this document be followed. Until results from research can fully explain the physical-chemical properties of CNT and CNF that define their inhalation toxicity, all types of CNT and CNF should be considered a respiratory hazard and exposure should be controlled below the REL.

In addition to controlling exposures below the REL, it is prudent for employers to institute medical surveillance and screening programs for workers who are exposed to CNT and CNF for the purpose of possibly detecting early signs of adverse pulmonary effects including fibrosis. Such an assessment can provide a secondary level of prevention should there be inadequacies in controlling workplace exposures. In 2009, NIOSH concluded that there was insufficient evidence to recommend specific medical tests for workers exposed to the broad category of engineered nanoparticles but when relevant toxicological information became available, specific medical screening recommendations would be forthcoming [NIOSH 2009b]. As described in this document, the toxicologic evidence on CNT/CNF has advanced to make specific recommendations for the medical surveillance and screening of exposed workers. That is, the strong evidence for pulmonary fibrosis from animal studies and the fact that this effect can be detected by medical tests is the basis for NIOSH specific medical screening recommendations. NIOSH also recommends other risk management practices in addition to controlling exposure and medical surveillance. These include education and training of workers and the use of personal protective equipment (e.g., respirators, clothing, and gloves).

In summary, the findings and recommendations in this Current Intelligence Bulletin are intended to minimize the potential health risks associated with occupational exposure to CNT and CNF by recommending a working lifetime exposure limit (1 μg/m^3, 8-hr TWA, 45 years), a sampling and analytical method to detect CNT and CNF, medical surveillance and screening and other guidelines. The expanding use of CNT/CNF products in commerce and research warrants these protective actions.

Background

The goal of this occupational safety and health guidance for carbon nanotubes (CNT) and carbon nanofibers (CNT) is to prevent the development of adverse respiratory health effects in workers. To date, NIOSH is not aware of any reports of adverse health effects in workers producing or using CNT or CNF. The concern about worker exposure to CNT or CNF arises from the results of recent laboratory animal studies with CNT and CNF. Shortterm and subchronic studies in rats and mice have shown qualitatively consistent noncancerous adverse lung effects including pulmonary inflammation, granulomas, and fibrosis with inhalation, intratracheal instillation, or pharyngeal aspiration of several types of CNT (single or multiwall; purified or unpurified). These earlystage, noncancerous adverse lung effects in animals include: (1) the early onset and persistence of pulmonary fibrosis in CNT-exposed mice [Shvedova et al., 2005, 2008; Porter et al., 2010; Mercer et al., 2011], (2) an equal or greater potency of CNT compared with other inhaled particles known to be hazardous (e.g., crystalline silica, asbestos) in causing pulmonary inflammation and fibrosis [Lam et al., 2004; Shvedova et al., 2005; Muller et al., 2005], and (3) reduced lung clearance in mice or rats exposed to relatively low-mass concentrations of CNT [Mercer et al., 2009; Pauluhn 2010a]. Findings of acute pulmonary inflammation and interstitial fibrosis have also been observed in mice exposed to CNF [Murray et al., 2012]. The extent to which these animal data may predict clinically significant lung effects in workers is not known. However, NIOSH considers these animal study findings of pulmonary inflammation, granulomas, and fibrosis associated with exposure to CNT and CNF to be relevant to human health risk assessment because similar lung effects have been observed in workers in dusty jobs [Rom and Markowitz 2006; Hubbs et al., 2011].

Some studies also indicate that CNT containing certain metals (nickel, 26%) [Lam et al., 2004] or higher metal content (17.7% vs. 0.2% iron) are more cytotoxic in vitro and in vivo [Shvedova et al., 2003, 2008]. However, in experimental animal studies, both unpurified and purified (low metal content) CNT are associated with early onset and persistent pulmonary fibrosis and other adverse lung effects [Lam et al., 2004; Shvedova et al., 2005; 2008]. Other studies indicate that differences in physical-chemical properties, including functionalization or bio-modification, may alter the lung retention and biological responses [Kagan et al., 2010; Osmond-McLeod et al., 2011; Pauluhn 2010a; Oyabu et al., 2011]. Although a number of different types of CNT and CNF have been evaluated, uncertainty exists on the generalizability of the current animal findings to new CNT and CNF.

In addition to the earlystage noncancer lung effects in animals, some studies in cells or animals have shown genotoxic or carcinogenic effects. In vitro studies with human lung cells have shown that singlewalled carbon nanotubes (SWCNT) can cause genotoxicity and abnormal chromosome number by interfering with mitosis (cell division) [Muller et al., 2008b; Sargent et al., 2009, 2011; Kisin et al., 2011]. Other in vitro studies did not show evidence of genotoxicity of some MWCNT [Wirnitzer et al., 2009; Kim et al., 2011].

Studies in mice exposed to multiwalled carbon nanotubes (MWCNT) have shown the migration of MWCNT from the pulmonary alveoli to the intrapleural space [Hubbs et al., 2009; Porter et al., 2010; Mercer et al., 2010]. The intrapleural space is the same site in which malignant mesothelioma can develop due to asbestos exposure. Intraperitoneal injection of CNT in mice has resulted in inflammation from long MWCNT (> 5 μm in length), but not short MWCNT (< 1 μm in length) or tangled CNT [Poland et al., 2008; Takagi et al., 2008;

Muller et al., 2009; Murphy et al., 2011]. In rats administered CNT by peritoneal injection, the pleural inflammation and mesothelioma were related to the thin diameter and rigid structure of MWCNT [Nagai et al., 2011]. In a study of rats administered MWCNT or crocidolite by intrapulmonary spraying, exposure to either material produced inflammation in the lungs and pleural cavity in addition to mesothelial proliferative lesions [Xu et al., 2012].

Pulmonary exposure to CNT has also produced systemic responses including an increase in inflammatory mediators in the blood, as well as oxidant stress in aortic tissue and increase plaque formation in an atherosclerotic mouse model [Li et al., 2007; Erdely et al., 2009]. Pulmonary exposure to MWCNT also depresses the ability of coronary arterioles to respond to dilators [Stapleton et al., 2011]. These cardiovascular effects may be due to neurogenic signals from sensory irritant receptors in the lung. Mechanisms, such as inflammatory signals or neurogenic pathways causing these systemic responses, are under investigation.

Additional research is needed to fully explain the mechanisms of biological responses to CNT and CNF, and the influence of physical-chemical properties. The findings of adverse respiratory effects and systemic effects reported in several animal studies indicate the need for protective measures to limit worker exposure to CNT and CNF.

CNT and CNF are currently used in many industrial and biomedical applications, including electronics, lithiumion batteries, solar cells, super capacitors, thermoplastics, polymer composites, coatings, adhesives, biosensors, enhanced electron-scanning microscopy imaging techniques, inks, and in pharmaceutical/biomedical devices. CNT and CNF can be encountered in facilities ranging from research laboratories and production plants to operations where CNT and CNF are processed, used, disposed, or recycled. The data on worker personal exposures to CNT and CNF are extremely limited, but reported workplace airborne concentrations for CNT [Maynard et al., 2004; Han et al., 2008a; Bello et al., 2009, 2010; Tsai et al., 2009; Lee et al., 2010; Cena and Peters 2011; Dahm et al., 2011] and CNF [Methner et al., 2007; Evans et al., 2010; Birch 2011a; Birch et al., 2011b] indicate the potential for worker exposures in many tasks or processes and the reduction or elimination of exposures when measures to control exposure are used.

Assessment of the Health Risk and Recommended Exposure Limit

NIOSH has determined that the best data to use for a quantitative risk assessment and as basis for a recommended exposure limit (REL) are the nonmalignant pulmonary data from the CNT animal studies. At present, data on cancer and cardiovascular effects are not adequate for a quantitative risk assessment of inhalation exposure. NIOSH considers the pulmonary responses of inflammation and fibrosis observed in shortterm and subchronic studies in animals to be relevant to humans, as inflammatory and fibrotic effects are also observed in occupational lung diseases associated with workplace exposures to other inhaled particles and fibers. Uncertainties include the extent to which these lung effects in animals are associated with functional deficits and whether similar effects would be clinically significant among workers. However, these fibrotic lung effects observed in some of the animal studies developed early (e.g., 28 days after exposure) in response to relatively low-mass lung doses, and also persisted or progressed after the end of exposure [Shvedova et al., 2005, 2008; MaHock et al., 2009; Pauluhn 2010a; Porter et al., 2010; Mercer et al., 2011; DeLorme et al., 2012; Murray et al., 2012]. Given the relevance of these types of lung effects to humans, the

REL was derived using the published subchronic and shortterm animal studies with doseresponse data of early stage fibrotic and inflammatory lung responses to CNT exposure (Section 5 and Appendix A).

Critical effect levels for the noncancerous lung effects estimated from the animal doseresponse data (e.g., BMD, benchmark dose and BMDL, the 95% lower confidence limit estimates of the BMD) have been extrapolated to humans by accounting for the factors influencing the lung dose in each animal species. The no observed adverse effect level (NOAEL) and lowest observed adverse effect level (LOAEL) estimates reported in the subchronic inhalation studies were also evaluated as the critical effect levels. Working-lifetime exposure concentrations were calculated based on estimates of either the deposited or retained alveolar lung dose of CNT assuming an 8-hour time-weighted average (TWA) exposure during a 40-hour workweek, 50 weeks per year, for 45 years. Based on BMD modeling of the subchronic animal inhalation studies with MWCNT [Ma-Hock et al., 2009; Pauluhn 2010a], a working lifetime exposure of 0.2–2 μg/m^3 (8-hour TWA concentration) was estimated to be associated with a 10% excess risk of earlystage adverse lung effects (95% lower confidence limit estimates) (Tables 5–1 and A–5). Risk estimates derived from shortterm animal studies (Tables A–3 and A–4) were consistent with these estimates.

In addition to the BMD-based risk estimates, NOAEL or LOAEL values were used as the critical effect level in animals. As with the BMD(L) estimates, the human-equivalent working lifetime concentrations were estimated, although using dosimetric adjustment and uncertainty factors (Section A.6.3). The estimated human-equivalent working lifetime concentrations based on this approach were approximately 4–18 μg/m^3 (8-hr TWA), depending on the subchronic study and the interspecies dose retention and normalization factors used. Dividing these estimates by data-suitable uncertainty factors (e.g., UFs of 20–60), and assuming a threshold model, the estimated zero risk levels were <1 μg/m^3 as working lifetime 8-hr TWA concentrations. A recent subchronic inhalation (13-wk exposure plus 3 months follow-up) study of CNF in rats [DeLorme et al., 2012] showed qualitatively similar lung response as in a shorter-term (28-day) study of CNF administered by pharyngeal aspiration in mice [Murray et al., 2012] (Sections 3.5 and A.7). Using the NOAEL-based approach, the human-equivalent working lifetime concentration estimates were 1–4 μg/m^3 (8-hr TWA), depending on the data and assumptions used to estimate the human-equivalent dose (Section A.7).

In the 2010 draft Current Intelligence Bulletin (CIB) *Occupational Exposure to Carbon Nanotubes and Nanofibers*, NIOSH proposed a REL of 7 μg/m^3 elemental carbon (EC) 8-hr TWA, which was set at the upper limit of quantitation (LOQ) for NIOSH Method 5040 [NIOSH 2010a]. In the draft CIB, NIOSH acknowledged that workers may still have an excess risk of developing earlystage pulmonary effects including fibrosis if exposed over a full working lifetime at the proposed REL. In view of these health risks, and ongoing improvements in sampling and analytical methodologies, NIOSH is recommending a REL of 1 μg/m^3 EC as an 8-hr TWA respirable mass concentration using NIOSH Method 5040 (Section 6.1, Appendix C). The 45-yr working lifetime excess risk estimates of minimal level (grade 1 or greater) lung effects in rats observed by histopathology at 1 μg/m^3 (8-hr TWA concentration) range from 2.4% to 33% (maximum likelihood estimates, MLE) and 5.3% to 54% (95% upper confidence limit, UCL) estimates (Table A–7). The 45-yr working lifetime excess risk estimates of slight/mild (grade 2) lung effects at 1 μg/m^3 (8-hr TWA) range from 0.23% to 10% MLE and 0.53% to 16% (95% UCL) (Tables 5–2 and A–8). These estimates are based on a risk assessment using dose-response data from the rat subchronic inhalation studies of two types of MWCNT. The range in these risk estimates reflects differences across studies and/or types of MWCNT and the uncertainty in the estimation of working lifetime CNT lung burden. The lung burden estimates are based on either the retained lung dose (normal clearance) or deposited lung dose (no clearance). Although data from animal studies with CNF are more limited [Murray et al., 2012; DeLorme et al., 2012], physical-chemical similarities between CNT and CNF and findings of acute pulmonary inflammation and interstitial fibrosis in animals exposed to CNF [Murray et al., 2012] indicate the need to also control occupational exposure to CNF at the REL of 1 μg/m^3 EC. Because of uncertainties in the risk estimates some residual risk for adverse lung effects may exist at the REL; therefore, efforts should be made to reduce airborne concentrations to CNT and CNF as low as possible. Until the results from animal research studies can fully explain the mechanisms (e.g., shape, size, chemistry, functionalized) that potentially increase or decrease their toxicity all types of CNT and CNF should be considered a respiratory hazard and occupational exposures controlled at the REL of 1 μg/m^3.

Exposure Measurement and Controls

Occupational exposure to all types of CNT and CNF can be quantified using NIOSH Method 5040. A multitiered exposure measurement strategy is recommended for determining worker exposure to CNT and CNF [Section 6.1]. When exposure to other types of EC (e.g., diesel soot, carbon black) are absent or negligible, environmental background EC concentrations are typically < 1 μg/m^3 including in facilities where CNT and CNF are produced and used [Evans et al., 2010; Birch 2011a, b; Dahm et al., 2011]. Thus, an elevated airborne EC concentration relative to background (environmental and in non-process areas in the workplace) is a reasonable indicator of CNT or CNF exposure. When exposure to other types of EC is possible, additional analytical techniques may be required to better characterize exposures. For example, analysis of airborne samples by transmission electron microscopy (TEM) equipped with energy dispersive x-ray spectroscopy (EDS) can help to verify the presence of CNT and CNF (Section 6.1.2).

Published reports of worker exposure to CNT and CNF using NIOSH Method 5040 (EC determination) are limited but in the study by Dahm et al. [2011] worker personal breathing zone (PBZ) samples collected at CNT manufacturers frequently found low to nondetectable mass concentrations of EC when engineering controls were present. In a study by Birch et al. [2011a], the outdoor air concentrations over four survey days, two months apart, were nearly identical, averaging about 0.5 μg/m^3. Respirable EC area concentrations inside the facility were about 6–68 times higher than outdoors, while personal breathing zone samples were up to 170 times higher. In studies where airborne particle concentrations were used as a surrogate for measuring the potential release of CNT and CNF, the use of engineering controls (e.g., local exhaust ventilation, wet cutting of composites, fume hood/ enclosures) appeared to be effective in reducing worker exposure [Han et al., 2008; Bello et al., 2009; Tsai et al., 2009; Methner et al., 2010a; Cena and Peters 2011] (Section 2.1). However, direct reading instruments used in these studies are non-selective tools and often subject to interferences due to other particle sources, especially at low concentrations [Evans et al., 2010; Birch et al., 2011]. Control strategies and technologies developed by several industrial trade associations have proven successful in managing micrometer-sized fine powder processes, and should have direct application to controlling worker exposures from CNT and CNF processes. Examples include guidance issued for containing dry powder during manufacturing of detergents by the Association Internationale de la Savonnerie, de la Détergence et des Produits d'Entretien (AISE) [AISE 2001]. Following these guidelines makes it possible, at a minimum, to control enzyme-containing dust exposures below 60 ng/m^3 for enzymes. Additional guidance on a broader process and facility approach is available from the International Society for Pharmaceutical Engineering (ISPE). This organization offers guidance on the design, containment, and testing of various processes that handle finely divided dry powder formulations. One guide in particular, Baseline Guide Volume 1, 2nd Edition: Active Pharmaceutical Ingredients Revision to Bulk Pharmaceutical Chemicals, has broad applicability to CNT and CNF processes and is available from ISPE [ISPE 2007]. Finally, the Institute for Polyacrylate Absorbents (IPA) has developed guidelines for its member companies to assist them in controlling worker exposures to fine polyacrylate polymer dust in the micrometer-size range through a combination of engineering controls and work practices [IPA 2013]. The extent to which worker exposure to CNT and CNF can be controlled below 1 µg/m^3 respirable mass concentration as an 8-hr TWA is unknown, but should be achievable in most manufacturing and end-use job tasks if engineering controls are used and workers are instructed in the safe handling of CNT/CNF materials.

Until results from research studies can fully explain the physical-chemical properties of CNT and CNF that define their inhalation toxicity, all types of CNT and CNF should be considered a respiratory hazard, and exposures should be controlled as low as possible below the REL. The REL is based on the respirable airborne mass concentration of CNT and CNF because the adverse lung effects in animals were observed in the alveolar (gasexchange) region. "Respirable" is defined as the aerodynamic size of particles that, when inhaled, are capable of depositing in the alveolar region of the lungs [ICRP 1994]. Sampling methods have been developed to estimate the airborne mass concentration of respirable particles [ACGIH 1984; CEN 1993; ISO 1995; NIOSH 1998]. Reliance on a respirable EC massbased REL will provide a means to identify job tasks with potential exposures to CNT and CNF so that appropriate measures can be taken to limit worker exposure.

Recommendations

In light of current scientific evidence from experimental animal studies concerning the hazard potential of CNT and CNF, steps should be taken to implement an occupational health surveillance program that includes elements of hazard and medical surveillance. NIOSH recommends that employers and workers take the following steps to minimize potential health risks associated with exposure to CNT and CNF.

1. Recommendations for Employers

- Use available information to continually assess current hazard potential related to CNT and CNF exposures in the workplace and make appropriate changes (e.g., sampling and analysis, exposure control) to protect worker health. At a minimum, follow requirements of the OSHA Hazard Communication Standard [CFR

1910.1200(h)] and the Hazardous Waste Operation and Emergency Response Standard [29 CFR 1910.120].

- Identify and characterize processes and job tasks where workers encounter bulk ("free-form") CNT or CNF and materials that contain CNT/CNF (e.g., composites).
- Substitute, when possible, a nonhazardous or less hazardous material for CNT and CNF. When substitution is not possible, use engineering controls as the primary method for minimizing worker exposure to CNT and CNF.
- Establish criteria and procedures for selecting, installing, and evaluating the performance of engineering controls to ensure proper operating conditions. Make sure workers are trained in how to check and use exposure controls (e.g., exhaust ventilation systems).
- Routinely evaluate airborne exposures to ensure that control measures are working properly and that worker exposures are being maintained below the NIOSH REL of 1 $\mu g/m^3$ using NIOSH Method 5040 (Section 6 and Appendix C).
- Follow exposure and hazard assessment procedures for determining the need for and selection of proper personal protective equipment, such as clothing, gloves, and respirators (Section 6).
- Educate workers on the sources and job tasks that may expose them to CNT and CNF, and train them about how to use appropriate controls, work practices, and personal protective equipment to minimize exposure (Section 6.3).
- Provide facilities for hand washing and encourage workers to make use of these facilities before eating, smoking, or leaving the worksite.
- Provide facilities for showering and changing clothes, with separate facilities for storage of nonwork clothing, to prevent the inadvertent cross-contamination of nonwork areas (including take-home contamination).
- Use light-colored gloves, lab coats, and workbench surfaces to make contamination by dark CNT and CNF easier to see.
- Develop and implement procedures to deal with cleanup of CNT and CNF spills and decontamination of surfaces.
- When respirators are provided for worker protection, the OSHA respiratory protection standard [29 CFR 1910.134] requires that a respiratory protection program be established that includes the following elements:
  - A medical evaluation of the worker's ability to perform the work while wearing a respirator.
  - Regular training of personnel.
  - Periodic workplace exposure monitoring.
  - Procedures for selecting respirators.
  - Respirator fit testing.
  - Respirator maintenance, inspection, cleaning, and storage.
- The voluntary use of respirators are permitted, but must comply with the provisions set forth in CFR 1910.134(c)(2)(i) and CFR 1910.134(c)(2)(ii).
- Information on the potential health risks and recommended risk management practices contained in this CIB should, at a minimum, be used when developing labels and Safety Data Sheets (SDS), as required [http://www.osha.gov/dsg/hazcom].

1.1. Medical Screening and Surveillance

The evidence summarized in this document leads to the conclusion that workers occupationally exposed to CNT and CNF may be at risk of adverse respiratory effects. These workers may benefit from inclusion in a medical screening program to help protect their health (Section 6.7).

1.1.1. Worker Participation

Workers who could receive the greatest benefit from medical screening include the following:

- Workers exposed to concentrations of CNT or CNF in excess of the REL (i.e., all workers exposed to airborne CNT or CNF at concentrations above 1 μg/m^3 EC as an 8-hr TWA).
- Workers in areas or jobs that have been qualitatively determined (by the person charged with program oversight) to have the potential for intermittent elevated airborne concentrations to CNT or CNF (i.e., workers are at risk of being exposed when they are involved in the transfer, weighing, blending, or mixing of bulk CNT or CNF, or the cutting or grinding of composite materials containing CNT or CNF, or workers in areas where such activities are carried out by others).

1.1.2. Program Oversight

Oversight of the medical surveillance program should be assigned to a qualified healthcare professional who is informed and knowledgeable about potential workplace exposures, routes of exposure, and potential health effects related to CNT and CNF.

1.1.3. Screening Elements

Initial Evaluation

- An initial (baseline) evaluation should be conducted by a qualified healthcare professional and should consist of the following:
  - An occupational and medical history, with respiratory symptoms assessed by use of a standardized questionnaire, such as the American Thoracic Society Respiratory Questionnaire [Ferris 1978] or the most recent.
  - A physical examination with an emphasis on the respiratory system.
  - A spirometry test (Anyone administering spirometry testing as part of the medical screening program should have completed a NIOSH-approved training course in spirometry or other equivalent training; additionally, the health professional overseeing the screening and surveillance program should be expert in interpreting spirometry testing results, enabling followup evaluation as needed.).
  - A baseline chest X-ray (digital or filmscreen radiograph). All baseline chest images should be clinically interpreted by a board eligible/certified radiologist or other physician with appropriate expertise, such as a board eligible/certified pulmonologist. Periodic follow up chest X-rays may be considered, but there is currently insufficient evidence to evaluate effectiveness. However, if periodic

follow up is obtained, clinical interpretation and classification of the images by a NIOSH-certified B reader using the standard International Classification of Radiographs of Pneumoconioses (ILO 2011 or the most recent equivalent) are recommended.

- Other examinations or medical tests deemed appropriate by the responsible healthcare professional (The need for specific medical tests may be based on factors such as abnormal findings on initial examination—for example, the findings of an unexplained abnormality on a chest X-ray should prompt further evaluation that might include the use of high-resolution computed tomography scan of the thorax.).

Periodic Evaluations

- Evaluations should be conducted at regular intervals and at other times (e.g., postincident) as deemed appropriate by the responsible healthcare professional based on data gathered in the initial evaluation, ongoing work history, changes in symptoms such as new, worsening, or persistent respiratory symptoms, and when process changes occur in the workplace (e.g., a change in how CNT or CNF are manufactured or used or an unintentional "spill"). Evaluations should include the following:
    - An occupational and medical history update, including a respiratory symptom update, and focused physical examination—performed annually.
    - Spirometry—testing less frequently than every 3 years is not recommended [OSHA NIOSH 2011]; and
    - Consideration of specific medical tests (e.g., chest X-ray).

Written reports of medical findings

- The healthcare professional should give each worker a written report containing the following:
    - The individual worker's medical examination results.
    - Medical opinions and/or recommendations concerning any relationships between the individual worker's medical conditions and occupational exposures, any special instructions on the individual's exposures and/or use of personal protective equipment, and any further evaluation or treatment.
- For each examined employee, the healthcare professional should give the employer a written report specifying the following:
    - Any work or exposure restrictions based on the results of medical evaluations.
    - Any recommendations concerning use of personal protective equipment.
    - A medical opinion about whether any of the worker's medical conditions is likely to have been caused or aggravated by occupational exposures.
- Findings from the medical evaluations having no bearing on the worker's ability to work with CNT or CNF should not be included in any reports to employers. Confidentiality of the worker's medical records should be enforced in accordance with all applicable regulations and guidelines.

1.1.4. Worker Education

Workers should be provided information sufficient to allow them to understand the nature of potential workplace exposures, potential health risks, routes of exposure, and instructions for reporting health symptoms. Workers should also be provided with information about the purposes of medical screening, the health benefits of the program, and the procedures involved.

1.1.5. Periodic Evaluation of Data and Screening Program

- Standardized medical screening data should be periodically aggregated and evaluated to identify worker health patterns that may be linked to work activities and practices that require additional primary prevention efforts. This analysis should be performed by a qualified health professional or other knowledgeable person to identify worker health patterns that may be linked to work activities or exposures. Confidentiality of workers' medical records should be enforced in accordance with all applicable regulations and guidelines.
- Employers should periodically evaluate the elements of the medical screening program to ensure that the program is consistent with current knowledge related to exposures and health effects associated with occupational exposure to CNT and CNF.

Other important components related to occupational health surveillance programs, including medical surveillance and screening, are discussed in Appendix B.

2. Recommendations for Workers

- Ask your supervisor for training in how to protect yourself from the potential hazards associated with your job, including exposure to CNT and CNF.
- Know and use the exposure control devices and work practices that keep CNT and CNF out of the air and off your skin.
- Understand when and how to wear a respirator and other personal protective equipment (such as gloves, clothing, eyewear) that your employer might provide.
- Avoid handling CNT and CNF in a 'free particle' state (e.g., powder form).
- Store CNT and CNF, whether suspended in liquids or in a powder form, in closed (tightly sealed) containers whenever possible.
- Clean work areas at the end of each work shift (at a minimum) using a HEPA-filtered vacuum cleaner or wet wiping methods. *Dry sweeping or air hoses should not be used to clean work areas.*
- Do not store or consume food or beverages in workplaces where bulk CNT or CNF, or where CNT-or CNF-containing materials, are handled.
- Prevent the inadvertent contamination of nonwork areas (including take-home contamination) by showering and changing into clean clothes at the end of each workday.

ABBREVIATIONS

AF	adjustment factor
ALvSA	Alveolar lung surface area
APS	aerodynamic particle sizer
APF	assigned protection factor
BEL	benchmark exposure limit
BET	Brunauer-Emmett-Teller
BMC	benchmark concentration (maximum likelihood estimate)
BMCL	95% lower confidence limit estimate of BMC
BMC(L)	Refers to both the BMC and BMCL estimates
BMD	benchmark dose (maximum likelihood estimate)
BMDL	95% lower confidence limit estimate of BMD
BMD(L)	Refers to both the BMD and BMDL estimates
BMDS	benchmark dose software
BMR	benchmark response
BW	body weight
C	carbon
CIB	Current Intelligence Bulletin
cm^2	centimeter squared
CNF	carbon nanofibers
CNM	carbon-based nanomaterial
CNT	carbon nanotubes
Co	cobalt
CPC	condensation particle counter
CVD	chemical vapor deposition
d	day
DAF	dosimetric adjustment factor
DF	deposition fraction
DPM	diesel particulate matter
EC	elemental carbon
EDS	energy dispersive spectroscopy
ELPI	electrical low pressure impactor
EPA	Environmental Protection Agency
ESP	electrostatic precipitator
Fe	iron
FMPS	fast mobility particle sizer
FPSS	fast particulate size spectrometer
g	grams
GM	geometric mean
GSD	geometric standard deviation
HCL	hydrogen chloride
HEC	human equivalent concentration
HEPA	high efficiency particulate air
hr	hour(s)

ISO	International Organization for Standardization
IT	intratracheal instillation
kg	kilogram
L	liters
LCL	lower confidence limit
LEV	local exhaust ventilation
LOAEL	lowest observed adverse effect level
LOD	limit of detection
LOQ	limit of quantitation
lpm	liters per minute
m^2	meter squared
mg	milligram(s)
mg/kg	milligram per kilogram body weight
mg/m^3	milligrams per cubic meter
m/s	meters per second
min	minute
ml	milliliters
MLE	maximum likelihood estimate
MMAD	mass median aerodynamic diameter
MPPD	multiple-path particle dosimetry
MWCNT	multi-walled carbon nanotubes
NEDO	New Energy and Industrial Technology Development Organization
NF	normalizing factor
Ni	nickel
NIOSH	National Institute for Occupational Safety and Health
nm	nanometer(s)
NOAEL	no observed adverse effect level
NR	not reported
NTRC	Nanotechnology Research Center
OC	organic carbon
OEL	occupational exposure limit
OSHA	Occupational Safety and Health Administration
PA	pharyngeal aspiration
PBZ	personal breathing zone
PMN	polymorphonuclear neutrophils
POD	point of departure
PPE	personal protective equipment
REL	recommended exposure limit
ROS	reactive oxygen species
RPD	relative percent difference
RSD	relative standard deviation
RT	retention half-time
SD	standard deviation
SDS	Safety Data Sheets
SEM	scanning electron microscopy
SMPS	scanning mobility particle sizer

SWCNT	single-walled carbon nanotubes
TC	total carbon
TD	toxicodynamic
TEM	transmission electron microscopy
TK	toxicokinetic
TP	thermal precipitator
TWA	time-weighted average
UF	uncertainty factor
μg	microgram(s)
$\mu g/m^3$	micrograms per cubic meter
μm	micrometer(s)
UCL	upper confidence limit
U.S.	United States
VE	ventilation rate
wk	week
Y	yttrium
yr	year
%	percent

ACKNOWLEDGMENTS

This Current Intelligence Bulletin (CIB) was developed by the scientists and staff of the National Institute for Occupational Safety and Health (NIOSH) who participate in the NIOSH Nanotechnology Research Center (NTRC). Paul Schulte is the manager, and Charles Geraci is the coordinator of the NIOSH National Occupational Research Agenda (NORA) nanotechnology cross-sector program. The document manager was Ralph Zumwalde. The document writing team was Ralph Zumwalde, Eileen Kuempel, Eileen Birch, Douglas Trout, and Vincent Castranova. The data analyses were performed by Eileen Kuempel, Randall Smith, Ralph Zumwalde, and Eileen Birch. Document internal review was conducted by Thomas Lentz, Faye Rice, John Bailer, David Dankovic, Christine Sofge, Leonid Turkevich, Douglas Evans, Ronald Shaffer, Mary Schubauer-Berigan, Aleksandr Stefaniak, Charles Geraci, and Paul Schulte.

Special thanks goes to NIOSH researchers who provided rodent dose-response data used in the risk assessment and contributed to the interpretation of the pulmonary response and aerosol data: Ann Hubbs, Robert Mercer, Elena Kisin, Anna Shvedova, Dale Porter, Bean Chen, and Pramod Kulkarni.

Special acknowledgement also goes to the NTRC scientists whose research cited in this CIB has significantly contributed to the scientific data used to support recommendations for minimizing worker exposure to carbon nanotubes and nanofibers.

The NIOSH NTRC acknowledges the contributions of Vanessa Williams and Rachel Thieman for graphic design and document production and John Lechliter for editing the CIB.

Acknowledgment is also due to Julie Muller, Ph.D., Dominique Lison, Ph.D., Lan Ma-Hock, Ph.D., Robert Landseidel, Ph.D., and Jürgen Pauluhn, Ph.D. for providing additional data and information from their published studies that were used in the risk assessment.

Peer Reviewers

The NIOSH NTRC greatly appreciates the time and effort of the expert peer reviewers who provided comments and suggestions on the December 2010 public and stakeholder draft of the CIB.

Robert Aitken, Ph.D
Director of Strategic Consulting, Institute of Occupational Medicine (IOM)
Nanotechnology Programme Director, IOM
Director of SAFENANO
Institute of Occupational Medicine
Riccarton, Edinburgh

Andrew D. Maynard, Ph.D
Charles and Rita Gelman Professor of Risk Science
Director, University of Michigan Risk Science Center

Philip Sayre, Ph.D
Associate Director, Risk Assessment Division
Office of Pollution Prevention & Toxics, U.S. EPA

Ronald H. White, M.S.T.
Deputy Director, Risk Sciences and Public Policy Institute
Johns Hopkins University
Bloomberg School of Public Health

Public Review

NIOSH greatly appreciates the public comments on the December 2010 draft document that were submitted to the NIOSH docket. The comments and responses to them can be seen at: www.cdc.gov/niosh/docket/archive/docket161A.html

1. Introduction

Many nanomaterial-based products are now commercially available. These include nanoscale powders, solutions, and suspensions of nanoscale materials, as well as composite materials and devices incorporating nanomaterials. The International Organization for Standardization (ISO) has developed nomenclature and terminology for nanomaterials [ISO/TS 2008]. According to ISO 27687:2008, a nano-object is material with one, two, or three external dimensions in the size range from approximately 1–100 nanometers (nm). Subcategories of a nano-object are (1) nanoplate, a nano-object with one external dimension at the nanoscale (i.e., 1–100 nm); (2) nanofiber, a nano-object with two external dimensions at the nanoscale, with a nanotube defined as a hollow nanofiber and a nanorod as a solid nanofiber; and (3) nanoparticle, a nano-object with all three external dimensions at the nanoscale. Nano-objects are commonly incorporated in a larger matrix or substrate called a nanomaterial. This taxonomy differs slightly from that suggested by the title of this CIB "*Occupational Exposure to Carbon Nanotubes and Nanofibers*." By this title, NIOSH is not suggesting an alternative taxonomy, but rather identifying the nano-objects (nanoscale carbon fiber and tube structures) that have been evaluated to date in toxicology and workplace exposure measurement studies.

Carbon nanotubes (CNT) are nanoscale cylinders of carbon (essentially consisting of seamlessly "rolled" sheets of graphene) that can be produced with very large aspect ratios. There is no single type of carbon nanotube. They may differ in shape, dimension, physical characteristics, surface coatings, chemical composition, or surface functionalization. This includes "raw" CNT, which contain residual metal catalysts vs. "purified" CNT, from which most of the metal catalysts have been removed. Singlewalled carbon nanotubes (SWCNT) consist of a single rolled graphene sheet and have a typical diameter of approximately 1–2 nm. Multiwalled carbon nanotubes (MWCNT) consist of many singlewalled tubes stacked one inside the other with diameters in the range of 2–100 nm, depending on the number of encapsulated tubes forming the CNT structure. SWCNT and MWCNT can vary in length, with some being up to many tens of micrometers long [Thostenson et al., 2001]. Carbon nanofibers (CNF), which are structurally similar to MWCNT, have typical diameters approximately 40 to 200 nm [Ku et al. [2006]. CNF have lengths ranging from tens of micrometers to several centimeters, average aspect ratios (length to diameter ratio) of > 100, and they display various morphologies, including cupped or stacked graphene structures. The primary characteristic that distinguishes CNF from CNT resides in graphene plane alignment. If the graphene plane and fiber axis do not align, the structure is defined as CNF, but when parallel, the structure is considered a CNT [ISO/TS 2008].

The synthesis of CNT and CNF requires a carbon source and an energy source [Sanchez et al., 2009]. CNT and CNF are synthesized by several distinct methods, including chemical vapor deposition (CVD), arc discharge, laser ablation, and highpressure CO conversion (HiPco). Depending on material and method of synthesis, a metal catalyst maybe used to increase yield and sample homogeneity, and to reduce the synthesis temperature. The diameter of the fibers depends on the dimensions of the metal nanoparticle used as a catalyst; the shape, symmetry, dimensions, growth rate, and crystallinity of the materials are influenced by the selection of the catalyst, carbon source, temperature, and time of the reaction. Different amounts of residual catalyst often exist following synthesis; consequently, post-synthesis treatments are used to increase the purity of the product. The most common purification technique involves selective oxidation of the amorphous carbon and/or carbon shells at a controlled temperature followed by washing or sonicating the material in an acid (HCL, HNO_3, H_2SO_4) or base (NaOH) to remove the catalyst. As there are many types of purification processes, purified CNT and CNF will exhibit differences in the content of trace elements and residual materials [Liu et al., 2008; Hou et al., 2008].

A growing body of literature indicates a potential health hazard to workers from exposure to various types of carbon nanotubes and nanofibers. A number of research studies with rodents have shown adverse lung effects at relatively low-mass doses of CNT (Tables 3–2 and 3–7), including pulmonary inflammation and rapidly developing, persistent fibrosis. Similar effects have been recently observed with exposure to CNF (Table 3–6). It is not known how universal these adverse effects are, that is, whether they occur in animals exposed to all types of CNT and CNF, and whether they occur in additional animal models. Most importantly, it is not yet known whether similar adverse health effects occur in humans following exposure to CNT or CNF, or how airborne CNT in the workplace may compare in size and structure to the CNT aerosols generated in the animal studies.

Because of their small size, structure, and low surface charge, CNT and CNF can be difficult to separate in the bulk form and tend to be agglomerated or to agglomerate quickly when released in the air, which can affect their potential to be inhaled and deposited in the lungs. The extent to which workers are exposed to CNT and CNF in the form of agglomerates or as single tubes or structures is unclear because of limited exposure measurement data, but airborne samples analyzed by electron microscopy have shown both individual and agglomerated structures [Johnson et al., 2010; Methner et al., 2010b; Birch et al., 2011b; Dahm et al., 2011].

This Current Intelligence Bulletin (CIB) summarizes the adverse respiratory health effects that have been observed in laboratory animal studies with SWCNT, MWCNT, and CNF. A recommended exposure limit (REL) for CNT and CNF is given to help minimize the risk of occupational respiratory disease in workers as well as guidance for the measurement and control of exposures to CNT and CNF.

2. Potential for Exposure

The novel application of CNT and CNF has been extensively researched because of their unique physical and chemical properties. CNT and CNF are mechanically strong, flexible, lightweight, heat resistant, and they have high electrical conductivity [Walters et al., 1999; Yu et al., 2000]. The commercial market for CNT and CNF is expected to grow substantially over the next decade [Lux Research 2007] with global capacity in 2013 estimated at 2,000 tons/year for MWCNT and 6 tons/year for SWCNT [Nanotech 2013]. Carbon nanotubes and nanofibers are commercially used in a variety of applications. These include electronics, lithiumion batteries, solar cells, super capacitors, thermoplastics, polymer composites, coatings, adhesives, biosensors, enhanced electron/scanning microscopy imaging techniques, and inks. They are also used in pharmaceutical/biomedical devices for bone graft ing, tissue repair, drug delivery, and medical diagnostics [WTEC 2007; Milne et al., 2008].

The potential for worker exposure to CNT and CNF can occur throughout the life cycle of CNT- and CNF-product use (processing, use, disposal, recycling) [Maynard and Kuempel 2005] (Figure 2–1), but the extent to which workers are exposed has not been completely characterized. Available data indicate that airborne exposures to CNT and CNF can occur during the transfer, weighing, blending, and mixing of the bulk powders, and during the cutting and drilling of CNT- and CNF-composite materials. A recent study of U.S. companies manufacturing carbonaceous nanomaterials identified 43 companies manufacturing CNT (14 primary, 18 secondary, and 11 primary and secondary users) [Schubauer-Berigan et al., 2011]. The purpose of the study was to enumerate the companies directly manufacturing (or using in other manufacturing processes) engineered carbonaceous nanomaterials in the United States, and to estimate the workforce size and characteristics of nanomaterials produced. The number of workers engaged in the manufacturing of CNT was estimated at 375, with a projected growth rate in employment of 15% to 17% annually.

The quantity of CNT (SWCNT and MWCNT) produced annually by each company was estimated to range from 0.2 to 2500 kg. The size of the workforce involved in the fabrication or handling of CNT/CNF-enabled materials and composites is unknown, but it is expected to increase as the market expands from research and development to industrial high-volume production [Invernizzi 2011].

2.1. Exposure to Carbon Nanotubes during Research and Development, Small-scale Manufacturing, and Use Applications

Recent assessments of airborne exposure to MWCNT in a research laboratory that manufactures and handles MWCNT found totalparticulate concentrations ranging from 37 $\mu g/m^3$ (weighing operation) to 430 $\mu g/m^3$ (blending process) in the absence of exposure controls [Han et al., 2008a].

The implementation of engineering controls (e.g., ventilated enclosure of MWCNT blending process) significantly reduced airborne particulate concentrations, often to non-detectable results. Transmission electron microscopy (TEM) analysis (NIOSH Method 7402) of personal breathing zone (PBZ) and area samples collected during the blending of MWCNT found airborne concentrations ranging from 172.9 tubes/cm^3 (area sample) to 193.6 tubes/cm^3 (PBZ sample) before the installation of exposure controls. The subsequent introduction of exposure controls significantly reduced airborne MWCNT concentrations to 0.018–0.05 tubes/cm^3. Aerosolized MWCNT structures were reported to have 52–56 nm diameters and 1473–1760 nm (avg. 1.5 μm) lengths.

Maynard et al. [2004] also assessed the propensity for SWCNT to be released during the agitation of unprocessed SWCNT material in a laboratorybased study and during the handling (e.g., furnace removal, powder transfer, cleaning) of unrefined material at four smallscale SWCNT manufacturing facilities in which laser ablation and highpressure carbon monoxide techniques were used to produce SWCNT.

Particle measurements taken during the agitation of unprocessed material in the laboratory indicated the initial airborne release of material (some visually apparent) with the particle concentration of the aerosol (particles < 0.5 μm in diameter) observed to decrease rapidly over time. With no agitation, particles around 0.1 μm in diameter appeared to be released from the SWCNT material, probably because of the airflow across the powder. At the four manufacturing facilities, shortterm SWCNT mass concentrations were estimated (using a catalyst metal as the surrogate measurement) to range from 0.7 to 53 $\mu g/m^3$ (area samples) in the absence of exposure controls. When samples were evaluated by scanning electron microscopy (SEM), most of the aerosolized SWCNT were agglomerated, with agglomerated sizes typically larger than 1 μm.

Airborne particle and MWCNT concentrations were determined by Bello et al. [2008] during chemical vapor deposition (CVD), growth, and handling of vertically aligned CNT films. Continuous airborne particle measurements were made using a real-time fast mobility particle sizer (FMPS) and a condensation particle counter (CPC) throughout the furnace operation. No increase in total airborne particle concentration (compared with background) was observed during the removal of MWCNT from the reactor furnace or during the detachment of MWCNT from the growth substrate (a process whereby MWCNT are removed from the substrate with a razor blade).

Electron microscopic analysis of a PBZ sample collected on the furnace operator found no detectable quantity of MWCNT, either as individual tubes or as agglomerates. No mention was made about the use of engineering controls (e.g., local exhaust ventilation, fume hood) to prevent exposure to MWCNT.

The potential for airborne particle and SWCNT and MWCNT release was determined in a laboratory setting in which both types of CNT were produced using CVD [Tsai et al., 2009]. A qualitative assessment of the exposure (i.e., particle morphology, aerosol size) was made during the synthesis of SWCNT and MWCNT in which modifications of the manufacturing methods were made to ascertain how changes in the production of CNT influenced airborne particle size and concentration (e.g., SWCNT synthesis with and without a catalyst, and growth of MWCNT on a substrate and with no substrate). An FMPS and an aerodynamic particle sizer (APS) were used to monitor particle size and concentrations. Background particle concentrations were determined to assist in quantifying the release of SWCNT and MWCNT during their synthesis and handling. Samples were also collected for analysis by TEM to determine particle morphology and elemental composition. Particle measurements made inside a fume hood during the synthesis of SWCNT were found to be as high as 10^7 particles/ cm^3 with an average particle diameter of 50 nm; PBZ samples collected on workers near the fume hood were considerably lower (< 2,000 particles/ cm^3). The difference between particle concentrations obtained during SWCNT growth using a catalyst and the control data (no catalyst) was small and was postulated to be a result of particles being released from the reactor walls of the furnace even when no SWCNT were being manufactured. Particle measurements made during the synthesis of MWCNT were found to peak at 4×10^6 particles/ cm^3 when measured inside the fume hood. Particle size ranged from 25 to 100 nm when a substrate was used for MWCNT growth and from 20 to 200 nm when no substrate was present.

Airborne particle concentrations and particle size were found to vary because of the temperature of the reactor, with higher particle concentrations and smaller particle sizes observed at higher temperatures. PBZ samples collected on workers near the fume hood during MWCNT synthesis had particle concentrations similar to background particle concentrations.

TEM analysis of MWCNT samples indicated the presence of individual particles as small as 20 nm with particle agglomerates as large as 300 nm. Some individual MWCNT were observed, but were often accompanied by clusters of carbon and iron particles. The diameters of the tubes were reported to be about 50 nm. The use of a fume hood that was extra wide and high and operated at a constant velocity of 0.7 m/s face velocity, appeared to be effective in minimizing the generation of turbulent airflow at the hood face, which contributed to the good performance of the fume hood in capturing the airborne release of SWCNT and MWCNT during their synthesis.

Lee et al. [2010] investigated the potential airborne release of MWCNT at seven facilities (e.g., research laboratories, small-scale manufacturing) where MWCNT was either being synthesized by CVD or handled (e.g., ultrasonic dispersion, spraying). Real-time aerosol monitoring was conducted using a scanning mobility particle sizer (SMPS) and a CPC to determine particle size and concentration. PBZ and area samples were collected for determining airborne mass concentrations (total suspended particulate matter) and for TEM (NIOSH Method 7402) and SEM analysis for particle identification and characterization. Background measurements of airborne nanoparticle exposures were determined at two of the seven worksites before starting work to assist in establishing a baseline for airborne nanoparticle concentrations. Most of the handling of MWCNT during synthesis and application was performed inside a laboratory fume hood, where most of the measurements were made.

Exposure concentrations of total suspended particulate matter ranged from 0.0078 to 0.3208 mg/m^3 for PBZ samples and 0.0126 to 0.1873 mg/m^3 for area samples. TEM and SEM analysis of filter samples found no detectable amounts of MWCNT but only aggregates of metal particles (e.g., iron and aluminum) which were used as catalysts in the synthesis of MWCNT. The highest airborne particle releases were observed in area samples collected during catalyst preparation (18,600–75,000 particles/cm^3 for 20–30 nm diameter particles) and during the opening of the CVD reactor (6,974–16,857 particles/cm^3 for 20–50 nm diameter particles). Other handling processes such as CNT preparation, ultrasonic dispersion, and opening the CNT spray cover also generated the release of nanoparticles. The ultrasonic dispersion of CNT generated particles in the range of 120 to 300 nm, which were larger in size than those released from other processes.

The release of airborne carbon-based nanomaterials (CNMs) was investigated during the transfer and ultrasonic dispersion of MWCNT (10–20 nm diameters), fullerenes, and carbon black (15 nm diameter) inside a laboratory fume hood with the airflow turned off and the sash halfway open [Johnson et al., 2010]. Airborne exposure measurements were made during the weighing and transferring of dry CNMs to beakers filled with reconstituted freshwater with and without natural organic matter and then sonicated. The study was designed to determine the relative magnitude of airborne nanomaterial emissions associated with tasks and materials used to evaluate environmentally relevant matrices (e.g., rivers, ponds, reservoirs). Direct reading real-time instruments (i.e., CPC, OPC) were used to determine airborne particle number concentrations, with the results compared with particle number concentrations determined from general air samples collected in the laboratory before and after the laboratory process. Samples were also collected for TEM analysis to verify the presence of CNMs.

Airborne particle number concentrations for all tasks exceeded background particle concentrations, which were inversely related to particle size, with the size distribution of particles skewed toward those CNMs with an aerodynamic diameter < 1 μm. Airborne particle number concentrations for MWCNT and carbon black, during the sonication of water samples, were significantly greater than those found during the weighing and transferring of dry CNMs. TEM analysis of airborne area samples revealed agglomerates of all CNMs, with MWCNT agglomerates observed to be 500 to 1,000 nm in diameter.

The National Institute for Occupational Safety and Health (NIOSH) conducted emission and exposure assessment studies at 12 sites where engineered nanomaterials were produced or used [Methner et al., 2010a]. Studies were conducted in research and development laboratories, pilot plants, and smallscale manufacturing facilities handling SWCNT, MWCNT, CNF, fullerenes, carbon nanopearls, metal oxides, electrospun nylon, and quantum dots. Airborne exposures were characterized using a variety of measurement techniques (e.g., CPC, OPC, TEM) [Methner et al., 2010b]. The purpose of the studies was to determine whether airborne exposures to these engineered nanomaterials occur and to assess the capabilities of various measurement techniques in quantifying exposures. In a research and development laboratory handling CNF, airborne particle number concentrations (determined by CPC) were reported as 4000 particles/ cm^3 during weighing/mixing and 5000 particles/

cm^3 during wet sawing. These concentrations were substantially less than the reported background particle concentration of 19,500 particles/cm^3. Samples collected for TEM particle characterization indicated the aerosol release of some CNF. All handling of CNF was in a laboratory hood (with HEPA filtered vacuum) for the weighing/mixing and wet saw cutting of CNF composite materials. In a facility making CNF in a chemical vapor phase reactor, OPC particle count concentrations ranged from 5,400 particles/cm^3 (300–500 nm particle size) to a high of 139,500 particles/cm^3 (500–100 nm particle size). Higher airborne particle concentrations were found during the manual scooping of CNF in the absence of exposure control measures. Samples collected for TEM particle characterization indicated the aerosol release of some CNF. In another research and development laboratory, the potential for airborne exposure to MWCNT was evaluated during weighing, mixing, and sonication. All handling of MWCNT was performed in a laboratory hood (without HEPA filtered vacuum). Particle concentrations were determined by CPC (particle size 10–1000nm) and OPC (particle size 300–500nm, 500–100nm). CPC particle concentrations ranged from 1480–1580 particles/cm^3 (weighing MWCNT in hood) to 2200–2800 particles/cm^3 (sonication of MWCNT).

The background particle concentration determined by CPC was 700 particles/cm^3. Airborne particle concentrations determined by OPC ranged from 3,900–123,400 particles/cm^3 (weighing) to 6,500–42,800 particles/ cm^3 (sonication). Background particle concentrations determined by OPC ranged from 700 particles/cm^3 (1–10 μm particle size) to 13,700 particles/cm^3 (300–500 nm particle size). The higher particle concentrations determined with the OPC indicated the presence of larger, possibly agglomerated particles. Samples collected for TEM particle characterization indicated the aerosol release of agglomerated MWCNT.

Subsequent studies conducted by NIOSH at six primary and secondary pilot or smallscale manufacturing facilities (SWCNT, MWCNT, CNF) employed a combination of filter-based samples to evaluate PBZ and area respirable and inhalable mass concentrations of EC as well as concentrations of CNT and CNF structures determined by TEM analysis [Dahm et al., 2011]. A total of 83 filter-based samples (30 samples at primary and 22 at secondary manufacturers) were collected for EC determination (NIOSH Method 5040) and 31 samples for TEM analysis (NIOSH Method 7402). Similar processes and tasks were reported in the three primary and three secondary manufacturers of CNT. These processes and tasks consisted of: (1) similar production and harvesting methods for CNT, and common cleaning/housekeeping procedures in primary manufacturers, and (2) common CNT handling practices such as weighing, mixing, sonication, manual transfer, cleaning, and spray coating operations in secondary manufacturers. Worker PBZ inhalable mass concentrations found at primary manufacturers ranged from 0.68 to 5.25 μg/m^3 EC with an average concentration of 2.42 μg/m^3. Area samples for EC determination from these samples ranged from non-detectable to 4.62 μg/m^3 while outdoor background samples ranged from non-detectable to 0.89 μg/m^3. [Note: these are inhalable mass fractions and may not be equivalent to the unmeasured respirable mass fractions. Thus, these concentrations cannot be compared directly to the NIOSH REL].

The highest airborne exposures were found during harvesting of CNT when no exposure control measures were used. In secondary manufacturers, PBZ inhalable mass concentrations for EC ranged from non-detectable to 7.86 μg/m^3 with area sample concentrations ranging from non-detectable to 2.76 μg/m^3. No EC was found in an outdoor background air sample.

The highest airborne exposures were found during extrusion and weighing of CNT when using a fume hood that was reported as not always in operation or being utilized properly during material handling. A majority of the reported EC mass concentrations in primary and secondary facilities were determined from airborne samples found to contain detectable amounts of EC between the LOD and LOQ of Method 5040.

Samples for TEM analysis (collected side-by-side with PBZ and area mass samples) reported CNT concentrations for PBZ samples ranging from non-detectable to 1.613 structures/cm^3 with the highest concentration for an area sample reported as 0.295 structures/cm^3. A statistical correlation between side-by-side mass concentrations and TEM structure counts was reported (p-0.01) with a corresponding Pearson correlation coefficient of 0.44. Various types of exposure prevention measures were reported for most workplaces including the use of PPE (e.g., respirators, gloves, safety glasses) and implementation of different exposure control techniques (e.g., glove box, chemical fume hood, clean rooms). A limitation of the study was that most workers were not handling CNT/CNF full shift, thus many samples were collected over a relatively short sample time due to the short duration of processes and tasks.

In a study designed to investigate the release of CNT during the dry and wet cutting of composite materials containing CNT, airborne samples were collected to determine particle number, respirable mass, and nanotube concentrations [Bello et al., 2009].

Two different composites containing MWCNT (10–20 nm diameters) were cut using a band saw or rotary cutting wheel. The laboratory study was designed to simulate the industrial cutting of CNT-based composites. PBZ and area samples (close to the emission source) were collected during dry cutting (without emission controls) and during wet cutting (equipped with a protective guard surrounding the rotary cutting wheel). The cutting of composite materials ranged from 1 to 3 minutes. The dry cutting of composite materials generated statistically significant ($P < 0.05$) quantities of airborne nanoscale and fine particles when compared with background airborne particle concentrations. Although the particle number concentration was dominated by the nanoscale and fine fractions, 71% to 89% of the total particle surface area was dominated by the respirable (1–10 μm) aerosol fraction. During the dry cutting of composites, reported mean PM10 mass concentrations for area samples were 2.11 and 8.38 mg/m^3, and 0.8 and 2.4 mg/m^3 for PBZ samples. Submicron and respirable fibers were generated from dry cutting of all composites. TEM analysis of area samples found concentrations that ranged from1.6 fibers/cm^3 (during the cutting of CNT-alumina) to 3.8 fibers/cm^3 (during the cutting of carbonbase composite materials). A PBZ fiber concentration of 0.2 fibers/cm^3 was observed during the dry cutting of base-alumina composite materials. No fiber measurement data were reported for the wet cutting of composite materials. No increase in mean PM10 mass concentrations were observed in 2 of 3 area samples collected during the wet cutting of composites. In the third sample, the observed high particle concentration was attributed to extensive damage of the protective guard around the rotary cutting wheel.

Bello et al. [2010] also investigated the airborne release of CNT and other nanosized fibers during solid core drilling of two types of advanced CNT-hybrid composites: (1) reinforced plastic hybrid laminates (alumina fibers and CNT), and (2) graphiteepoxy composites (carbon fibers and CNT). Worker PBZ and area samples were collected to determine exposures during the drilling of composite materials with local exhaust ventilation turned off. Four potential exposure-modifying factors were assessed: (1) by composite type,

(2) drilling rpm (low and high), (3) thickness of the composite, and (4) dry versus wet drilling. Replicate test measurements (10–30 measurements) lasting < 5 minutes were performed on each composite material. A combination of real-time and integrated samples were collected at the source and PBZ using an FMPS, aerodynamic particle sizer (APS), CPC, diffusion charger, and cascade impactor to measure aerosol particle size, concentration, and chemical identification. An electrostatic precipitator (ESP) and thermal precipitator (TP) were used to collect particles directly on TEM grids for electron microscopy analysis. Aerosol concentrations during high rpm drilling generally were higher than for low rpm drilling for all composite materials with aerosol concentrations found to be higher from alumina composites. Wet drilling was observed to suppress the release of particles > 10 nm in diameter. High aspect-ratio fiber concentrations were determined using the sizing and counting criteria in NIOSH Method 7400 (> 5 µm long, aspect ratio > 3). Airborne exposure to both alumina fiber and CNT structures were found ranging in concentration from 1.0 fibers/cm^3 (alumina composite) to 1.9 fibers/cm^3 (carbon and CNT composite) for PBZ samples; similar concentrations were observed in area samples. Because sampling volume and fiber surface density on the samples were below the optimal specification range of Method 7400, fiber concentration values were determined to be first order approximations. The authors concluded that higher input energies (e.g., higher drilling rpms, larger drill bits) and longer drill times associated with thicker composites generally produced higher exposures, and that the drilling of CNT-based composites generated a higher frequency of nanofibers than had been previously observed during the cutting of CNT-based composites [Bello et al., 2009].

Cena and Peters [2011] evaluated the airborne release of CNT during the weighing of bulk CNT and the sanding of epoxy nanocomposite sticks measuring 12.5 × 1.3 × 0.5 cm. Epoxy reinforced test samples were produced using MWCNT (Baytubes®) with 10–50 nm outer diameters and 1–20 µm lengths. The purpose of the study was to (1) characterize airborne particles during handling of bulk CNT and the mechanical processing of CNT composites, and (2) evaluate the effectiveness of local exhaust ventilation (LEV) hoods to capture airborne particles generated by sanding CNT composites. Airborne particle number and respirable mass concentrations were measured using a CPC (particle diameters 0.01 to 1 µm) and OPC (particle diameters 0.3 to 20 µm). Respirable mass concentrations were estimated using the OPC data. Samples for TEM analysis were also collected for particle and CNT characterization. PBZ and source airborne concentrations were determined during two processes: weighing bulk CNT and sanding epoxy nanocomposite test sticks. Exposure measurements were taken under three LEV conditions (no LEV, a custom fume hood, and a biological safety cabinet). CPC and OPC particle concentrations were measured inside a glove box in which bulk CNT (600 mg) was transferred between two 50-ml beakers; background particle concentrations were measured inside the glove box before the process began. To study the sanding process, a worker manually sanded test sticks that contained 2% by weight CNT. Aerosol concentrations were measured for 15–20 min in the worker's breathing zone and at a site adjacent to the sanding process. The sanding process with no LEV was conducted on a 1.2 m by 2.2 m worktable. The sanding was also conducted inside a custom fume hood that consisted of a simple vented enclosure that allowed airflow along all sides of the back panel but had no front sash or rear baffles. The average face velocity of the fume hood was 76 ft/min. Exposures from the sanding process were also assessed while using a biological safety cabinet (class II type A2).

Particle number concentrations determined during the weighing process contributed little to that observed in background samples (process to background ratio [P/B] = 1.06), however it did influence the mass concentration (P/B = 1.79). The GM respirable mass concentration inside the glove box was reported as 0.03 μg/m^3 (background GM was 0.02 μg/m^3). During the sanding process (including no LEV, in a fume hood, and in a biological safety cabinet) the PBZ nanoparticle number concentrations were negligible compared with background concentrations (P/B ratio = avg. 1.04). Particles generated during sanding were reported to be predominantly micron sized with protruding CNT and very different from bulk CNT that tended to remain in large (>1 μm) tangled agglomerates. Respirable mass concentrations in the worker's breathing zone were elevated. However, the concentrations were lower when sanding was performed in the biological safety cabinet (GM = 0.2 μg/m^3) compared with those with no LEV (GM was 2.68 μg/m^3) or those when sanding was performed inside the fume hood (GM = 21.4 μg/m^3; p value <0.0001). The poor performance of the fume hood was attributed to the lack of a front sash and rear baffles and its low face velocity.

2.2. Exposure to Carbon Nanotubes (Other Sources)

Exposure to CNT and other carbon nanocrystalline structures (e.g., spheres, shells) can occur during the burning of natural gas, propane, and other methane-series gases. CNT exposures have been observed in indoors areas where gas stoves are in use, and in outdoor air near gas-burning industrial sites [Murr et al., 2004 a, b]. Samples collected near a methane (CH_4)/air flame exhaust revealed carbon nanocrystal aggregates that ranged from 0.4 to 2 μm, and contained several thousand individual MWCNT and other nanocrystal (polyhedral) structures that averaged 20 nm in diameter.

2.3. Exposure to Carbon Nanofibers during Research and Small-Scale Manufacturing Operations

Some research has been conducted to date on workplace exposure to carbon nanofibers (CNF) [Methner et al., 2007; Evans et al., 2010; Birch 2011a; Birch et al., 2011b]. In a NIOSH health hazard evaluation conducted at a university-based research laboratory, the potential release of airborne CNF was observed at various processes using real-time aerosol instruments (e.g., CPC, ELPI, aerosol photometer) [Methner et al., 2007]. General area exposure measurements indicated slight increases in airborne particle number and mass concentrations relative to background measurements (outdoors and offices) during the transfer of CNF prior to weighing and mixing, and during the chopping and wet saw cutting of a polymer composite material. Airborne total carbon mass concentrations (per NIOSH Method 5040, with correction for adsorbed vapor) within the laboratory processing area were 2 to 64 times higher than those of a nearby office area, with the highest peak exposure concentration (1094 μg/m^3) found during the wet saw cutting of the CNF composite material. No indoor particle concentrations exceeded the outdoor background concentrations. Particles having a diameter of about 400 nm or greater were found in greater number during wet-saw cutting, while the number of particles having a diameter of about 500 nm or greater were elevated during the weighing and mixing of CNF. Airborne samples collected directly on

TEM grids were analyzed for the presence of CNF. Some fibers observed by TEM had diameters larger than the 100 nm criterion used to define a nanofiber, which was consistent with results reported by Ku et al. [2006], in which the mobility diameter of aerosolized CNF was observed to be larger than 60 nm, with a modal aerodynamic diameter of about 700 nm. The majority of CNF observed by TEM were loosely agglomerated, rather than single fibers, which was in general agreement with the particle size measurements made by real-time instruments.

Detailed investigations of exposures at different job tasks were conducted at a facility manufacturing and processing CNF [Evans et al., 2010; Birch 2011a; Birch et al., 2011b] in which CNF production totaled 14,000 kg a year. Filter, sorbent, cascade impactor, bulk, and microscopy samples, combined with direct-reading instruments, provided complementary information on air contaminants. Samples were analyzed for organic, elemental, and total carbon (OC + EC = TC), metals, and polycyclic aromatic hydrocarbons (PAHs), with EC as a measure of CNF. TEM with energy-dispersive X-ray spectroscopy (TEM-EDS) also was applied. Study results for the time-integrated (filter/sorbent) samples were reported in companion papers, with OC-EC, metals, and microscopy results in part I [Birch et al., 2011b] and PAH results in part II [Birch 2011a]. Fine/ultrafine, iron-rich soot, PAHs, and carbon monoxide were production byproducts. Area concentrations of respirable EC inside the facility were about 6 to 68 times higher than outdoors, while PBZ samples were up to 170 times higher.

As part of the same study, multiple direct-reading instruments (e.g., CPC, ELPI, photometer/with cyclone, diffusion charger, fast particulate size spectrometer [FPSS]) and respirable particle mass concentrations were used to assess CNF exposures on site and to evaluate instrument performance. A transient increase in respirable mass concentration was observed during manual bagging of the final product and was attributed to aerosolized CNF. The tamping of the bag to settle contents and the subsequent closing dispersed CNF through the bag opening into the workplace. High particle number and active surface area concentrations were found during the opening of the dryer and during the manual redistribution of the CNF product. This was attributed to the presence of ultrafine particles emitted from the dryer and as byproducts formed through the high-temperature thermal processing of CNF. No elevations in respirable mass concentrations were observed during these operations, suggesting that significant quantities of CNF were not released into the workplace. However, the transfer or dumping of dried CNF from a dryer to a drum, and subsequent bag change-out of final product, contributed the largest transient increases in respirable mass concentrations, with concentrations exceeding 1.1 mg/m^3 for transfer or dumping and 0.5 mg/m^3 for bag change-out. The authors concluded that integrated particle number and active surface area concentrations (i.e., using CPC and diffusion charger) were not useful in assessing the contribution of emissions from CNF in the workplace, because measurements were dominated by ultrafine particle emissions [Evans et al., 2010]. Respirable particle mass concentrations estimated by the photometer appeared to be the most useful and practical metric for measuring CNF when using direct-reading instruments. Results obtained for filter samples support the direct-reading instrument findings. The TEM analyses of size-selective area samples indicated that large fiber bundles were present [Birch et al., 2011b]. In addition, size-classified samples (collected with impactors) analyzed for EC (by NIOSH 5040) indicated CNF particles in the micrometer range [Birch et al., 2011b].

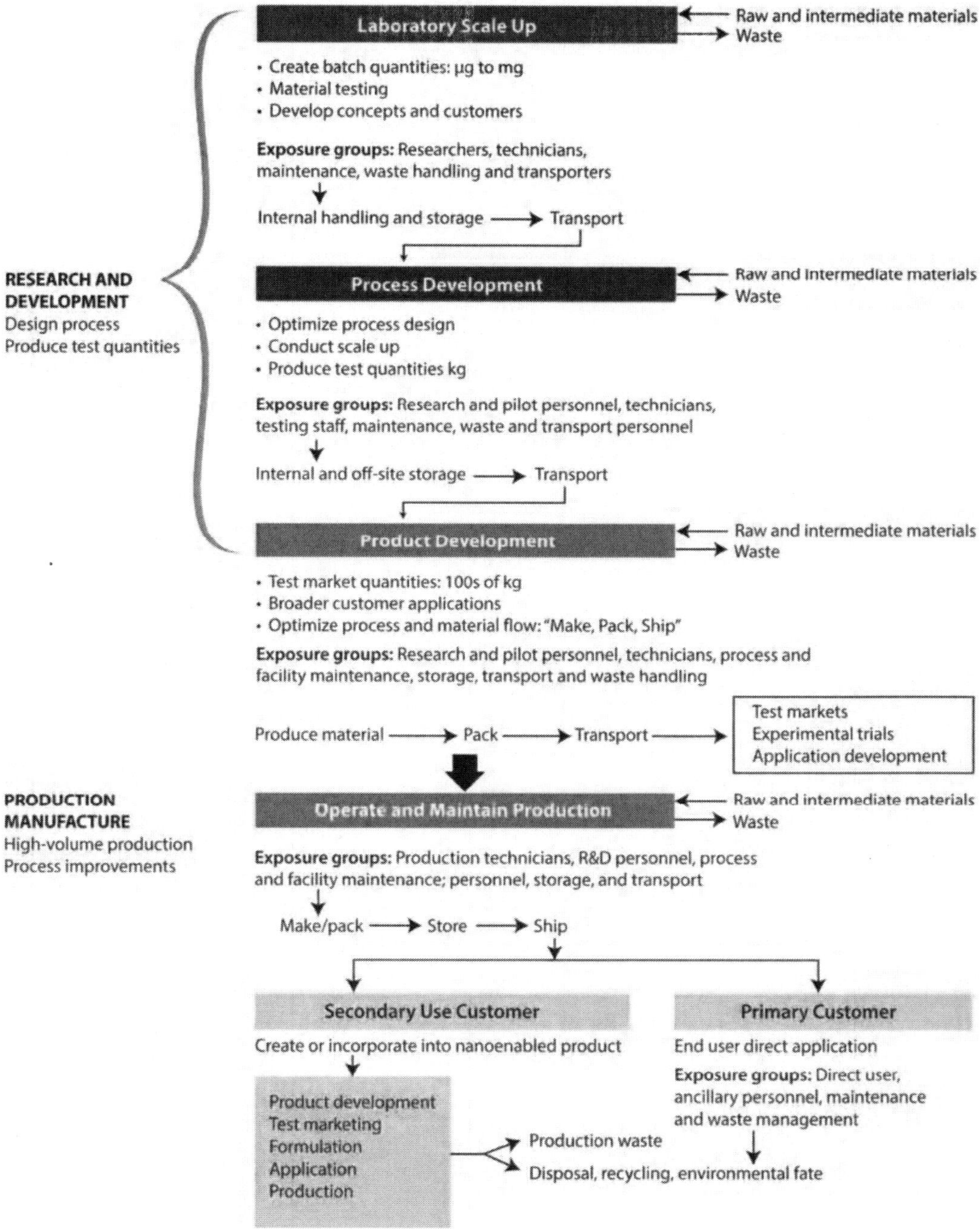

Figure 2–1. Workplaces and job tasks with potential for occupational exposure to carbon nanotubes and nanofibers. Adapted from Schulte et al., 2008.

3. Evidence for Potential Adverse Health Effects

Various types of laboratory animal studies have been conducted with CNT and CNF using different routes of exposure to evaluate potential toxicity (Tables 3–1 through 3–8). These studies have shown a consistent toxicological response (e.g., pulmonary inflammation, fibrosis) independent of the study design (i.e., intratracheal, aspiration, and inhalation). Exposure to SWCNT, MWCNT, and CNF are of special concern because of their small size and fibrous structure. The nanometer diameters and micrometer lengths of some materials closely resemble the dimensions of some mineral fibers (e.g., asbestos). Results from laboratory animal studies with SWCNT, MWCNT, and CNF also show pulmonary responses similar to those reported for some respirable particles and durable fibers. In some studies, CNT-induced lung fibrosis developed more rapidly and at a lower mass burden than either ultrafine carbon black or quartz [Lam et al., 2004; Shvedova et al., 2005; Ryman-Rasmussen et al., 2009b]. Pulmonary exposure to CNT has also produced systemic responses including an increase in inflammatory mediators in the blood, as well as oxidant stress in aortic tissue and increase plaque formation in an atherosclerotic mouse model [Li et al., 2007; Erdely et al., 2009; Stapleton et al., 2011].

CNT and CNF are widely considered durable in biological systems because of the process they undergo during synthesis in which contaminating catalytic metals are frequently removed by high temperature vaporization or acid/base treatment [Sanchez et al., 2009]. Researchers [Osmond- McLeod et al., 2011] have measured the durability (*in vitro*) of four types of CNT, one type of glass wool fiber, and two asbestos fiber types in simulated biological fluid (Gambles solution) followed by an assessment of their ability to induce an inflammatory response when injected into the abdominal cavities of mice. Three of the four types of CNT tested for durability showed no signs, or minimal loss of mass, or change in fiber length, or morphology. When the four CNTs were injected into the peritoneal cavity of mice, an inflammatory and fibrotic response was induced for three of the CNTs that had retained their long discrete structures, whereas the other CNT, which was less durable and had shorter structures and/or formed tight bundles, caused minimal inflammation. The findings were consistent with those found for asbestos and glass wool fibers after intraperitoneal injection into mice, in which an inflammatory and fibrotic response was elicited by the two asbestos samples that were both durable and contained a high percentage of long fibers, and the glass wool fiber that was not very durable and caused only minimal inflammation.

The physical-chemical properties (e.g., dimension, composition, surface characteristics) of CNT and CNF can often be modified to accommodate their intended commercial use. CNT and CNF can also be coated or functionalized, thus changing their surface chemistry. Toxicological effects of such changes remain largely unexplored [Yan et al., 2011] except for some limited evidence indicating that structural defects [Muller et al., 2008a; Fenoglio et al., 2008], surface [Sayes et al., 2006] and oxidative modification [Allen et al., 2008], nitrogen doping [Carrero-Sanchez et al., 2006], surface functionalization [Liu et al., 2010], and polymer (acid- and polystyrene-based) surface coating [Tabet et al., 2011] of CNT can influence their toxicity potential. Recent studies indicate that functionalization of MWCNT with –COOH groups significantly decreases the inflammatory and fibrotic response after aspiration in a mouse model [Sager et al., 2011; Wang et al., 2012] and when SWCNT were functionalized by carboxylation and subjected to phagolysosomal fluid, longitudinal splitting and oxidative degradation of the tubes occurred [Liu et al., 2010]. Kagan et al. [2010] reported that *in vitro* myeloperoxidase, which is found in high concentrations in polymorphonuclear neutrophils (PMN), degraded SWCNT. However, it is uncertain as to whether PMN-derived myeloperoxidase would degrade SWCNT *in vivo* (e.g., in the lung) because of the following: (1) PMN recruitment after SWCNT exposure is a transient rather than persistent response, (2) there is no strong evidence for SWCNT phagocytosis by PMN, and (3) SWCNT and MWCNT are found in the lungs of mice months after pharyngeal aspiration [Shvedova et al., 2005; Mercer et al., 2008; Porter et al., 2010].

Several animal studies have shown that the size (e.g., length) of MWCNT and SWCNT may have an effect on their biological activity [Takagi et al., 2008; Poland et al., 2008; Muller et al., 2009]. Intraperitoneal injection of mice with long MWCNT (20 μm length), but not short MWCNT (< 5 μm length), caused granulomatous lesions on the diaphragm in a 2-week post-exposure study [Poland et al., 2008]. Fibrotic peritoneal adhesions and mesothelioma were also observed after exposure to MWCNT in which approximately 28% of the tubes were > 5 μm in length [Takagi et al., 2008]. However, when rats were exposed to short MWCNT (< 1 μm length) by intraperitoneal injection, only acute inflammation was observed, with no evidence of mesothelioma over the 2 year post-exposure period [Muller et al., 2009].

Nagai et al. [2011] provided evidence that the carcinogenic potential of MWCNT may be related to the fiber-like properties and dimensions. Fischer 344/Brown Norway (male and female, 6 wk old) were injected with doses of 1 or 10 mg of one of five types of MWCNT with different dimensions and rigidity. The thin diameter MWCNT (~50 nm) with high crystallinity caused inflammation and mesothelioma, whereas thick (~150 nm) or tangled structures (~2–20 nm) were less cytotoxic, inflammogenic, or carcinogenic. A specific mutation to tumor suppressor genes (*Cdkn2a/2b*) was observed in the mesotheliomas, which is similar to that observed in asbestos-associated mesotheliomas induced by asbestos. *In vitro* studies with mesothelial cells showed that the thin MWCNT pierced cell membranes and caused cytotoxicity.

Numerous studies have investigated the genotoxic properties of CNT with results from *in vitro* assays indicating that exposure to SWCNT and MWCNT can induce DNA damage, micronuclei formation, disruption of the mitotic spindle, and induction of polyploidy [Li et al., 2005; Kisin et al., 2007; Muller et al., 2008a; Pacurari et al., 2008; Lindberg et al., 2009; Sargent et al., 2009; Asakura et al., 2010]. Other *in vitro* studies of some MWCNT did not show evidence of genotoxicity [Wirnitzer et al., 2009; Kim et al., 2011]. The presence of residual metal catalysts was also found to promote the generation of reactive oxygen species (ROS), thereby enhancing the potential for DNA damage [Pulskamp et al., 2007; Barillet et al., 2010]. The results from *in vitro* studies with CNF have also shown that exposure can cause genotoxicity [Magrez et al., 2006; Lindberg et al., 2009; Kisin et al., 2011] including aneugenic as well as clastogenic events. In addition, low-dose, longterm exposure of bronchial epithelial cells to SWCNT or MWCNT has been reported to transform these cells to exhibit unregulated proliferation, loss of contact inhibition of division, enhanced migration and invasion, and growth in solf agar [Stueckle et al., 20011]. When SWCNT-transformed epithelial cells were subcutaneously injected into the hind flanks of immunodeficient nude mice, small tumors were observed at one week post-injection. Histological evaluation of tumors showed classic cancer cell morphology, including the presence of multinucleated cells, an indicator of mitotic dysfunction [Wang et al., 2011].

When CNT and CNF are suspended in test media, agglomerates of various sizes frequently occur. This is particularly evident in test media used in recent studies where animals have been exposed to CNT suspensions by intratracheal instillation, intraperitoneal injection, or by pharyngeal aspiration (a technique where particle deposition closely resembles inhalation). The agglomerate size for CNT and CNF is normally smaller in a dry aerosol than when suspended in physiological media. Evidence from toxicity studies in laboratory animals indicates that decreasing agglomerate size increases the pulmonary response to exposure [Shvedova et al., 2007, 2008; Mercer et al., 2008]. The extent to which agglomerates of CNT and CNF de-agglomerate in biological systems (e.g., in the lung) is unknown. However, a diluted alveolar lining fluid has been shown to substantially improve dispersion of CNT in physiological saline [Porter et al., 2008; Wang et al., 2010a].

3.1. Single-Walled Carbon Nanotubes (SWCNT)

Mice or rats exposed to SWCNT by IT or pharyngeal aspiration have developed granulomatous lesions at sites in the lung where agglomerates of SWCNT deposited [Lam et al., 2004; Warheit et al., 2004]. In addition, interstitial fibrosis has also been reported [Shvedova et al., 2005; Mangum et al., 2006]. This fibrotic response was associated with the migration of smaller SWCNT structures into the interstitium of alveolar septa [Mercer et al., 2008].

3.1.1. IT Studies

Lam et al. [2004] investigated the toxicity of SWCNT obtained from three different sources, each with different amounts of residual catalytic metals being present. Mice were exposed by IT to three different types of SWCNT (containing either 27% Fe, 2% Fe, or 26% Ni and 5% Y) at concentrations of 0.1 or 0.5 mg and to carbon black (0.5 mg) or to quartz (0.5 mg). The mice were toxicologically assessed 7 or 90 days post exposure. All types of SWCNT studied produced persistent epithelioid granulomas (which were associated with particle agglomerates) and interstitial inflammation that were dose-related. No granulomas were observed in mice exposed to carbon black, and only mild to moderate inflammation of the lungs was observed in the quartz exposure group. High mortality (5/9 mice) occurred within 4 to 7 days in mice instilled with the 0.5 mg dose of SWCNT containing nickel and yttrium.

Warheit et al. [2004] exposed rats via IT to concentrations of 1 or 5 mg/kg SWCNT, quartz, carbonyl iron, or graphite particles, and evaluated effects at 24-hr, 1-week, 1-month, and 3-months post exposure. The SWCNT were reported to have nominal diameters of 1.4 nm and lengths > 1 µm, which tended to agglomerate into micrometer size structures. In this study, ~15% of the SWCNT-instilled rats died within 24 hours of SWCNT exposure, apparently due to SWCNT blockage of the upper air- ways. In the remaining rats, a transient inflammatory response of the lung (observed up to 1-month post exposure) and non-dose dependent multifocal granulomas that were non-uniform in distribution were observed. Only rats exposed to quartz developed a dose-dependent lung inflammatory response that persisted through 3 months. Exposures to carbonyl iron or graphite particles produced no significant adverse effects.

3.1.2. Pharyngeal Aspiration Studies

Progressive interstitial fibrosis of alveolar walls has also been reported in mice when exposed via pharyngeal aspiration to purified SWCNT at doses of 10, 20, 40 μg/mouse [Shvedova et al., 2005]. As with studies by Lam et al. [2004] and Warheit et al. [2004], epithelioid granulomas were associated with the deposition of SWCNT agglomerates in the terminal bronchioles and proximal alveoli. This granuloma formation was rapid (within 7 days), dose-dependent, and it persisted over the 60-day post exposure period. A rapid, dose-dependent, and progressive development of interstitial fibrosis in pulmonary regions distant from deposition sites of SWCNT agglomerates was observed, and it appeared to be associated with deposition of more dispersed SWCNT structures. At equivalent mass lung burdens, nano-sized carbon black failed to cause any significant pulmonary responses. These findings were consistent with those reported by Mangum et al. [2006], in which rats exposed to 2 mg/kg via pharyngeal aspiration developed granulomas at sites of SWCNT agglomerates and diffuse interstitial fibrosis at 21 days post exposure.

Also noted was the formation of CNT structures bridging alveolar macrophages, which may affect normal cell division and/or function. When a more dispersed delivery of SWCNT was given by aspiration to mice (10 μg) [Mercer et al., 2008], an accelerated increase in collagen production in the alveolar interstitium occurred that progressed in the absence of persistent inflammation, with the development of few granulomatous lesions. A significant submicrometer fraction of the dispersed SWCNT was observed to rapidly migrate into alveolar interstitial spaces with relatively little of the material being a target for macrophage engulfment and phagocytosis.

3.1.3. Inhalation Studies

Shvedova et al. [2008] compared the responses resulting from exposure via pharyngeal aspiration [Shvedova et al., 2005] with exposure via inhalation of more-dispersed SWCNT [Baron et al., 2008].

One set of mice were exposed by inhalation to 5 mg/m^3, 5 hr/day for 4 days, while mice exposed by aspiration were given a single dose of 10 or 20 μg. The SWCNT for both studies had dimensions of 0.8–1.2 nm diameters and 100–1000 nm lengths with a measured surface area (Brunauer-Emmett-Teller method [BET]) of 508 m^2/g. Both studies reported acute lung inflammation followed by the development of granulomatous pneumonia and persistent interstitial fibrosis; these effects were observed for both purified (0.2% Fe) [Shvedova et al., 2005] and unpurified (17.7% Fe) [Shvedova et al., 2008] SWCNT. The finding that the acute lung inflammation resolved after the end of exposure while the pulmonary fibrotic response persisted or progressed is unusual compared with lung responses observed from other inhaled particles.

The findings indicate that the mechanism may involve the direct stimulation of fibroblasts by dispersed SWCNT that translocate to the lung interstitium [Wang et al., 2010a, b]. Quantitatively, mice exposed by inhalation (dispersed SWCNT) were 4-fold more prone to developing an inflammatory response, interstitial collagen deposition, and fibrosis, when compared (at an estimated equivalent lung dose) with mice exposed by aspiration to a less dispersed suspension of SWCNT. The exposure of mice by inhalation of 5 mg/m^3 SWCNT [Shvedova et al., 2008] is relevant, because the Occupational Safety and Health Administration (OSHA) permissible exposure limit (PEL) for respirable synthetic graphite of 5 mg/m^3 is sometimes used for controlling workplace exposures to CNT.

3.2. Multi-Walled Carbon Nanotubes (MWCNT)

3.2.1. Pharyngeal Aspiration Studies

Exposures to well-dispersed MWCNT in mice via pharyngeal aspiration have resulted in dose- and time-dependent pulmonary inflammation [Han et al., 2008b; Wolfarth et al., 2009; Hubbs et al., 2009; Han et al., 2010; Porter et al., 2010; Mercer et al., 2011], as well as central nervous system effects [Sriram et al., 2007; Sriram et al., 2009], at doses ranging from 10 to 80 μg/mouse. Exposure of mice to dispersed suspension of purified MWCNT at doses of 10, 20, 40, or 80 μg resulted in pulmonary inflammation and damage, granulomas, and a rapid and persistent fibrotic response [Porter et al., 2010]. Morphometric analyses indicated that the interstitial fibrotic response was dose-dependent and progressed through 56 days post-exposure [Mercer et al., 2011]. There was also evidence that MWCNT can reach the pleura [Porter et al., 2010] and that alveolar macrophages containing MWCNT can migrate to the lymphatics and cause lymphatic inflammation [Hubbs et al., 2009]. Some of the MWCNT (mean diameter of 49 nm and mean length of 4.2 μm) were observed penetrating the outer lung wall and entered the intrapleural space [Hubbs et al., 2009; Mercer et al., 2010]. Morphometric analyses indicated that 12,000 MWCNT entered the intrapleural space at 56 days post-exposure to 80 μg of MWCNT [Mercer et al., 2010].

3.2.2. IT Studies

Lung inflammation and fibrosis have also been observed in rats exposed by IT to long (5.9 μm) or short (0.7 μm) MWCNT at doses of 0.5, 2, or 5 mg of either ground or unground MWCNT and examined up to 60 days post-exposure [Muller et al., 2005]. Rats that received ground MWCNT (0.7 μm) showed greater dispersion in the lungs, and fibrotic lesions were observed in the deep lungs (alveolar region). In rats treated with unground MWCNT (5.9 μm), fibrosis appeared mainly in the airways rather than in the lung parenchyma. The biopersistence of the unground MWCNT was greater than that of the ground MWCNT (81% vs. 36 %). At an equal mass dose, ground MWCNT produced a similar inflammatory and fibrogenic response as chrysotile asbestos and a greater response than ultrafine carbon black [Muller et al., 2005]. Similar findings have been reported by Aiso et al. [2010], in which rats exposed to IT doses of 0.04 and 0.16 mg of dispersed MWCNT (mean length-5 μm, diameter-88 nm) caused transient inflammation, and persistent granulomas and alveolar wall fibrosis. These acute effects have also been reported in guinea pigs at IT doses of 12.5 mg [Grubek-Jaworska et al., 2005] and 15 mg [Huczko et al., 2005]; in mice at doses of 0.05 mg (average diameter of 50 nm, average length of 10 μm) [Li et al., 2007], and at 5, 20, and 50 mg/kg [Park et al., 2009]; and in rats [Liu et al., 2008] dosed at 1, 3, 5, or 7 mg (diameters of 40 to 60 nm, lengths of 0.5 to 5 μm). In contrast, Elgrabli et al. [2008a] reported cell death but no histopathological lesions or fibrosis in rats exposed at doses of 1, 10, or 100 μg MWCNT (diameters of 20 to 50 nm, lengths of 0.5 to 2 μm). Likewise, Kobayashi et al. [2010] observed only transient lung inflammation and a granulomatous response in rats exposed to a dispersed suspension of MWCNT (0.04–1 mg/kg). No fibrosis was reported, but the authors did not use a collagen stain for histopathology, which would have compromised the sensitivity and specificity of their lung tissue analysis.

In a study of rats administered MWCNT or crocidolite asbestos by intrapulmonary spraying (IPS), exposure to either material produced inflammation in the lungs and pleural cavity in addition to mesothelial proliferative lesions [Xu et al., 2012]. Four groups of six rats each were given 0.5 ml of 500 μg suspensions, once every other day, five times over a 9-day period and then evaluated. MWCNT and crocidolite were found to translocate from the lung to the pleural cavity after administration. MWCNT and crocidolite were also observed in the mediastinal lymph nodes suggesting that a probable route of translocation of the fibers is lymphatic flow. Analysis of tissue sections found MWCNT and crocidolite in focal granulomatous lesions in the alveoli and in alveolar macrophages.

3.2.3. Inhalation Studies

Several shortterm inhalation studies using mice or rats have been conducted to assess the pulmonary [Mitchell et al., 2007; Arkema 2008; Ma-Hock et al., 2009; Porter et al., 2009; Ryman-Rasmussen et al., 2009b; Pauluhn 2010a; Wolfarth et al., 2011] and systemic immune effects [Mitchell et al., 2007] from exposure to MWCNT. Mitchell et al. [2007] reported the results of a whole-body shortterm inhalation study with mice exposed to MWCNT (diameters of 10 to 20 nm, lengths of 5 to 15 μm) at concentrations of 0.3, 1, or 5 mg/m^3 for 7 or 14 days (6 hr/day) (although there was some question regarding whether these structures were actually MWCNT [Lison and Muller 2008]). Histopathology of lungs of exposed animals showed alveolar macrophages containing black particles; however, there was no observed inflammation or tissue damage. Systemic immunosuppression was observed after 14 days, although without a clear concentration-response relationship. Mitchell et al. [2009] reported that the immunosuppression mechanism of MWCNT appears to involve a signal originating in the lungs that activates cyclooxygenase enzymes in the spleen. Porter et al. [2009] reported significant pulmonary inflammation and damage in mice 1 day after inhalation of well-dispersed MWCNT (10 mg/m^3, 5 hr/day, 2–12 days; mass aerodynamic diameter of 1.3 μm, count aerodynamic diameter of 0.4 μm). In addition, granulomas were also observed encapsulating MWCNT in the terminal bronchial/proximal alveolar region of the lung. In an inhalation (nose-only) study with mice exposed to 30 mg/m^3 MWCNT (lengths of 0.5 to 50 μm) for 6 hours, a high incidence (9 of 10 mice) of fibrotic lesions occurred [Ryman-Rasmussen et al., 2009b]. MWCNT were found in the subpleural region of the lung 1 day post exposure, with subpleural fibrosis occurring at 2 weeks post exposure that progressed through 6 weeks of followup. No fibrosis was observed in mice exposed to 1 mg/m^3 of MWCNT or in mice exposed to 30 mg/m^3 of nanoscale carbon black.

Subchronic inhalation studies with MWCNT have also been conducted in laboratory studies with rats to assess the potential dose-response and time course for developing pulmonary effects [Arkema 2008; Ma-Hock et al., 2009; Pauluhn 2010a]. Ma-Hock et al. [2009] reported on the results of a 90- day inhalation (head-nose) study with rats exposed at concentrations of 0.1, 0.5, or 2.5 mg/m^3 MWCNT (BASF Nanocyl NC 7000) for 6 hr/day, 5 days/week for 13 weeks with a resultant lung burden of 47–1170 μg/rat. No systemic toxicity was observed, but the exposure caused hyperplastic responses in the nasal cavity and upper airways (larynx and trachea), and granulomatous inflammation in the lung and in lung-associated lymph nodes at all exposure concentrations. The incidence and severity of the effects were concentration-related. No lung fibrosis was observed but pronounced alveolar lipoproteinosis did occur.

Ellinger-Ziegelbauer and Pauluhn [2009] conducted a shortterm inhalation bioassay (before the Pauluhn 2010a subchronic study) to investigate the dependence of pulmonary inflammation resulting from exposure to one type of MWCNT (Bayer Baytubes®), which was highly agglomerated and contained a small amount of cobalt (residual catalyst). Groups of rats were exposed to 11 mg/m^3 MWCNT containing either 0.53% or 0.12% cobalt to assess differences in pulmonary toxicity because of metal contamination. Another group of rats was exposed to 241 mg/m^3 MWCNT (0.53% cobalt) to serve the purpose of hazard identification. All animals were exposed to a single nose-only inhalation exposure of 6 hr followed by a post-exposure period of 3 months. Time course of MWCNT- related pulmonary toxicity was compared with rats exposed to quartz in post-exposure weeks 1, 4, and 13 to distinguish early, possibly surface area/activity-related effects from retention-related poorly soluble particle effects. Rats exposed to either quartz or MWCNT resulted in somewhat similar patterns of concentration-dependent pulmonary inflammation during the early phase of the study. The pulmonary inflammation induced by quartz increased during the 3 months post-exposure period, whereas that induced by MWCNT regressed in a concentration-dependent manner. The time course of pulmonary inflammation associated with retained MWCNT was independent on the concentration of residual cobalt. Pauluhn [2010a], using the same MWCNT (0.53% cobalt) used in the study by Ellinger-Ziegelbauer and Pauluhn [2009] exposed rats (nose-only) at concentrations 0.1, 0.4, 1.5, and 6 mg/m^3 for 6 hr/day, 5 days/week for 13 weeks. The aerosolized MWCNT were described as being highly agglomerated (mean diameter of 3 μm). Lung clearance of MWCNT at the low doses was slow, with a marked inhibition of clearance at 1.5 and 6 mg/m^3. Histopathology analysis at 6 months post exposure revealed exposure-related lesions in the upper respiratory (e.g., goblet cell hypermetaplasia and/or metaplasia) and lower respiratory (e.g., inflammation in the bronchiole-alveolar region) tract in animals exposed at concentrations of 0.4, 1.5, and 6 mg/m^3, as well as inflammatory changes in the distal nasal cavities that were similar to those found by Ma-Hock et al. [2009]. In rats exposed at 6 mg/m^3, a time-dependent increase of bronchioloalveolar hyperplasia was observed, as well as changes in granulomas and an increase in collagen deposition that persisted through the 39-week post-exposure observation period. No treatment-related effects were reported for rats exposed at 0.1 mg/m^3.

In a report submitted by Arkema [2008] to EPA, rats exposed (nose only) to agglomerates of MWCNT (Arkema) at concentrations of 0.1, 0.5, and 2.5 mg/ m^3 for 6 hr/day for 5 days exhibited histopathological effects that were consistent with those reported by Ma-Hock et al. [2009], Ellinger-Ziegelbauer and Pauluhn [2009] and Pauluhn [2010a]. An increase of various cytokines and chemokines in the lung, along with the development of granulomas were found in the 0.5 and 2.5 mg/m^3 exposure groups, while no treatment-related effects were reported at 0.1 mg/m^3.

3.3. SWCNT and MWCNT Intraperitoneal Studies

Intraperitoneal injection studies in rodents have been frequently used as screening assays for potential mesotheliogenic activity in humans. To date, exposures to only a few fiber types are known to produce mesotheliomas in humans; these include the asbestos minerals and erionite fibers. Several animal studies [Takagi et al., 2008; Poland et al., 2008; Muller et al., 2009; Varga and Szendi 2010; Murphy et al., 2011] have been conducted to investigate the hazard potential of various sizes and doses of MWCNT and SWCNT to cause a carcinogenic response. Takagi et al. [2008] reported on the intraperitoneal injection of 3 mg of MWCNT in p53 +/- mice (a tumor-sensitive, genetically engineered mouse model), in which approximately 28% of the structures were > 5 µm in length with an average diameter of 100 nm. After 25 weeks, 88% of mice treated with MWCNT revealed moderate to severe fibrotic peritoneal adhesions, fibrotic peritoneal thickening, and a high incidence of macroscopic peritoneal tumors. Histological examination found mesothelial lesions near fibrosis and granulomas. Similar findings were also seen in the crocidolite asbestos-treated positive control mice. Minimal mesothelial reactions and no mesotheliomas were produced by the same dose of (nonfibrous) C_{60} fullerene. Poland et al. [2008] reported that the peritoneal (abdominal) injection of long MWCNT—but not short MWCNT—induced inflammation and granulomatous lesions on the abdominal side of the diaphragm at 1 week post-exposure. This study, in contrast to the Takagi et al. [2008] study, used wild type mice exposed to a much lower dose (50 µg) of MWCNT. Although this study documented acute inflammation, it did not evaluate whether this inflammation would persist and progress to mesothelioma. Murphy et al. [2011] found similar findings in C57Bl/6 mice that were injected with different types of MWCNT composed of different tube dimensions and characteristics (e.g., tangled) or injected with mixed-length amosite asbestos. Mice were injected with a 5 µg dose directly into the pleural space and evaluated after 24 hours, 1, 4, 12, and 24 weeks. Mice injected with long (> 15 µm) MWCNT or asbestos showed significantly increased granulocytes in the pleural lavage, compared with the vehicle control at 24 hours post exposure. Long MWCNT caused rapid inflammation and persistent inflammation, fibrotic lesions, and mesothelial cell proliferation at the parietal pleural surface at 24 weeks post exposure. Short (< 4 µm) and tangled MWCNT did not cause a persistent inflammatory response and were mostly cleared from the intrapleural space within 24 hours.

A lack of a carcinogenic response was reported by Muller et al. [2009] and Varga and Szendi [2010] in rats, and by Liang et al. [2010] in mice, following intraperitoneal injection or implantation of MWCNT or SWCNT. No mesotheliomas were noted 2 years after intraperitoneal injection of MWCNT in rats at a single dose of 2 or 20 mg [Muller et al., 2009] or MWCNT (phosphorylcho- line-grafted) in mice when given daily doses of either 10, 50, or 250 mg/kg and evaluated at day 28 [Liang et al., 2010]. However, the MWCNT samples used in the Muller et al. [2009] and Liang et al. [2010] studies were very short (avg. < 1 µm in length observed by Muller et al. [2009] and < 2 µm in length observed by Liang et al. [2010]), and the findings were consistent with the low biological activity observed in the Poland et al. [2008] study when mice were exposed to short MWCNT. Varga and Szendi [2010] reported on the implantation of either MWCNT or SWCNT in F-344 rats (six per group) at a dose of 10 mg (25 mg/kg bw). Gelatin capsules containing either SWCNT (< 2 nm diameters × 4–15 µm lengths), MWCNT (10–30 nm diameters × 1–2 µm lengths), or crystalline zinc oxide (negative control) were implanted into the peritoneal cavity. Histological examination at 12 months revealed only a granulomatous reaction of foreign body type with epithelioid and multinucleated giant cells in CNT-exposed animals. No information was reported on what effect the delivery of SWCNT and MWCNT in gelatin capsules had on their dispersion in the peritoneal given the tendency of CNT to agglomerate. If SWCNT and MWCNT remained agglomerated following delivery, this may have resulted in the lack of a mesothelioma-inducing effect. The low biological activity observed for the short MWCNT sample (≤ 2 µm) used in the study was consistent with the findings from Poland et al. [2008], Muller et al. [2009], and Liang et al. [2010], in which short MWCNT were also used.

3.4. Systemic Responses to Pulmonary Exposure to SWCNT and MWCNT

Li et al. [2007] reported that multiple aspirations of SWCNT (20 μg/mouse, every 2 weeks, for 2 months) in Apo E -/- mice caused a 71% increase in aortic plaques. Inhalation of MWCNT by rats (26 mg/m^3 for 5 hr; lung burden of 22 μg) resulted in a 92% depression of the responsiveness of coronary arterioles to dilators 24 hr post-exposure [Stapleton et al., 2011], while aspiration of MWCNT has been shown to increase baroreflex activity in rats [Legramante et al., 2009; Coppeta et al., 2007]. Further-more, pharyngeal aspiration of MWCNT (80 μg/ mouse) results in induction of mRNA for certain inflammatory mediators and markers of blood/ brain barrier damage in the olfactory bulb, frontal cortex, midbrain and hippocampus brain regions 24 hr post-exposure [Sriram et al., 2009]. Several mechanisms have been suggested to explain these systemic responses:

3.4.1. Translocation of CNT to Systemic Sites

Translocation of intraperitoneal instilled MWCNT from the abdominal cavity to the lung has been reported [Liang et al., 2010]; however, there is no evidence that the reported systemic effects are associated with translocation of CNT from the lung to the affected tissue. Aspirated gold-labeled SWCNT were not found in any organ 2 weeks post-exposure [Mercer et al., 2009].

3.4.2. Systemic Inflammation

Pulmonary exposure to particles causes localized inflammation at the sites of particle deposition in the alveoli. Erdely et al. [2009] reported that aspiration of SWCNT or MWCNT (40 μg/m) induced a small but significant increase in blood neutrophils, mRNA expression, and protein levels for certain inflammatory markers in the blood at 4 hr post-exposure, but not at later times. Pulmonary CNT exposure also significantly elevated gene expression for mediators, such as Hif - 3α and S100α, in the heart and aorta at 4 hr post-exposure. Evidence also exists that pulmonary exposure to particles alters systemic micro-vascular function by potentiating PMN as they flow through pulmonary capillaries in the close proximity to affected alveoli. These potentiated blood PMN adhere to micro-vessel walls and release reactive species that scavenge NO produced by endothelial cells [Nurkiewicz et al., 2006; Nurkiewicz et al., 2009]. Therefore, less dilator-induced NO diffuses to vascular smooth muscle resulting in less dilation.

Table 3–1. Findings from an uncharacterized carbon nanotube short-term intratracheal instillation (IT) toxicology study

Study design and exposure/dose				Observed pulmonary effects		
Author/ year	**Species**	**Exposure route**	**Exposure or dose**	**Granuloma**	**Inflammation**	**Fibrosis**
Huczko et al. [2001]	G. pigs	IT of soot containing CNT (uncharacterized)	25 mg (eval: 28 days post exposure)	NR	–	NR

NR: Not Reported.

+ = effect observed.

– = no effect observed.

Table 3–2. Findings from published SWCNT short-term intratracheal instillation (IT) toxicology studies

Study design and exposure/dose				Observed pulmonary effects		
Author/year	**Species**	**Exposure route**	**Exposure or dose**	**Granuloma**	**Inflammation**	**Fibrosis**
Warheit et al. [2004]	Rats	IT	1, 5 mg (eval: 24-hr, 1-wk, 1 and 3 mo. post exposure)	+ non-dose dependent	transient	–
Lam et al. [2004]	Mice	IT	0.1, 0.5 mg (eval: 7 or 90 days post exposure]	+	+	NR
Inoue et al., [2008]	Mice	IT	4 mg (eval: 24-hr post exposure)	NR	+	NR

NR: Not Reported.
+ = effect observed.
– = no effect observed.

3.5. Carbon Nanofibers (CNF)

Recent observations indicate that exposure to CNF can cause respiratory effects similar to those observed in animals exposed to CNT [Murray et al., 2012]. In this study, female mice were exposed by pharyngeal aspiration to SWCNT (40 μg), CNF (120 μg) or crocidolite (120 μg) and evaluated post exposure at 1, 7, and 28 days. Delivered structure number or particle surface area at the highest doses were 1.89×10^6 and 0.042 m^2 for SWCNT, 4.14×10^6 and 0.05 m^2 for CNF, and 660×10^6 and 0.001 m^2 for asbestos. SWCNT and CNF were purified and contained 0.23 and 1.4% iron, respectively compared to the 18% iron of the asbestos sample. SWCNT had diameters of 1 to 4 nm and lengths ranging from 1 to 3 μm whereas the diameters of CNF ranged from 60 to 150 nm and lengths approximately 5 to 30 μm. The fiber lengths of asbestos ranged from 2 to 30 μm. On a mass dose bases, inflammation and lung damage at 1 day post-exposure followed the potency sequence of SWCNT>CNF>asbestos.

The same potency sequence was observed for TNF and IL-6 production at 1 day post-exposure. SWCNT agglomerates were associated with the rapid (7 days) development of granulomas, while neither CNF nor asbestos (being more dispersed) caused granulomatous lesions. Interstitial fibrosis (noted as TGF production, lung collagen, and Sirius red staining of the alveolar septa) was observed at 28 days post-exposure with a mass-based potency sequence of SWCNT>CNF=asbestos. The potency sequence for fibrosis was not found to be related to structure number or particle surface area (determined by BET gas absorption method) delivered to the lung. However, it is likely that gas absorption overestimates the surface area of agglomerated SWCNT structures delivered to the lung. Estimates of effective surface area, based on geometrical analysis of structures including agglomeration, provided an improved dose metric that was correlated to the toxicological responses to CNT and CNF.

Table 3–3. Findings from published SWCNT short-term aspiration toxicology studies

Study design and exposure/dose				Observed pulmonary effects		
Author/ year	**Species**	**Exposure route**	**Exposure or dose**	**Granuloma**	**Inflammation**	**Fibrosis**
Shvedova et al. [2005]	Mice	Pharyngeal aspiration	10, 20, 40 μg (eval: 1, 3, 7, 28, and 60 days post exposure)	+	+	+
Mangum et al. [2006]	Rats	Pharyngeal aspiration	2 mg/kg (eval: 1 or 21 days post exposure)	+	–	+ (interstitia l lesions)
Shvedova et al. [2007]	Mice (vitamin E deficient)	Pharyngeal aspiration	40 μg (eval: 1, 7, and 28 days post exposure)	+	+	+
Mercer et al. [2008]	Mice	Pharyngeal aspiration	10 μg (eval: 1-hr, 1 and 7 days and 1 mo. post exposure)	+ (undispersed) – (dispersed)	+	+
Shvedova et al. [2008]	Mice	Pharyngeal aspiration	5,10, 20 μg (eval: 1, 7, and 28 days post exposure)	+	+	+

NR=Not Reported.
+ = effect observed.
– = no effect observed.

Table 3–4. Findings from published SWCNT and CNF short-term inhalation toxicology studies

Study design and exposure/dose				Observed pulmonary effects		
Author/ year	Speci es	Exposure route	Exposure or dose	Granuloma	Inflammation	Fibrosis
Shvedova et al. [2008]	Mice	Inhalation	SWNCT—5 $mg/m^3$5 hr/day for 4 days (eval: 1, 7, and 28 days post exposure)	+	+	+
DeLorme et al. [2012]	Rats	Nose-only inhalation	CNF—0.54, 2.5 or 25 mg/m^3 6 hr/day for 90 days. (eval: 1 and 90 days post exposure)	–	–(0.54 mg/m^3) +(2.5 and 25 mg/m^3)	–

NR = Not Reported.
+ = effect observed.
– = no effect observed.

Table 3–5. Findings from published MWCNT short-term intratracheal (IT) instillation and intrapulmonary spraying toxicology studies

Study design and exposure/dose				Observed pulmonary effects		
Author/ year	Species	Exposure route	Exposure or dose	Granuloma	Inflammation	Fibrosis
Muller et al. [2005]	Rats	IT	0.5, 2, 5 mg (eval: 1 hr, 3, 15, 28, and 60 days post exp)	+	+	+
Huczko et al. [2005]	G. pigs	IT	15 mg (eval: 90 days post exp)	NT (pneumonia-like reaction)	+ (Increased lung resistance)	+/–
Grubek-Jaworska et al. [2005]	G. Pigs	IT	12.5 mg (eval: 90 days post exp)	+	+	+
Carrero-Sanchez et al. [2006]	Mice	IT	1, 2.5, 5 mg/kg (eval: 1, 2, 3, 7 and 30 days post exp)	+	+	+
Deng et al. [2007]	Mice	IT	600 µg (eval: 1 day post exp)	NR	–	NR

Table 3–5. (Continued)

Study design and exposure/dose				Observed pulmonary effects		
Author/ year	**Species**	**Exposure route**	**Exposure or dose**	**Granuloma**	**Inflammation**	**Fibrosis**
Li et al. [2007]	Mice	IT	0.05 mg (eval: 8, 16, and 24 days post exp)	NR	+	NR
Liu et al. [2008]	Rats	IT	1, 3, 5, and 7 mg/kg (eval: 1 and 7 days, 1 and 3 mo. post exp)	+	+	NR
Muller et al. [2008a]	Rats	IT	2 mg (eval: 3 and 60 days post exp)	+	+	NR
Muller et al. [2008b]	Rats	IT	0.5 or 2 mg (eval: 3 days post exp)	NR	+	NR
Inoue et al. [2008]	Mice	IT	4 mg/kg (eval: 1 day post exp)	NR	+	NR
Elgrabli et al. [2008a]	Rats	IT	1, 10, 100 µg (eval: 1, 7, 30, 90 and 180 d post exp)	–	–	–
Park et al. [2009]	Mice	IT	0.04, 0.2 or 1 mg/kg (eval: 3 days, 1 week, 1-mo, 3-mo, 6-mo post exp)	+	+	NR
Aiso et al. [2010]	Rats	IT	5, 20, or 50 mg/kg (eval: 1, 3, 7 or 14 days post exp)	+	transient	+
Kobayashi et al. [2010]	Rats	IT	0.04 or 0.16 mg (eval: 1, 7, 28 or 91 days post exp)	transient	transient	–
Xu et al. [2012]	Rats	Intrapulmonary	MWCNT—0.5 ml of 500 ug suspensions, 5 times over 9-days and then evaluated	+	+	NA

NR = Not Reported.
+ = effect observed.
– = no effect observed.

Table 3–6. Findings from published MWCNT or CNF short-term aspiration toxicology studies

Study design and exposure/dose				Observed pulmonary effects		
Author/year	**Species**	**Exposure route**	**Exposure or dose**	**Granuloma**	**Inflammation**	**Fibrosis**
Sriram et al. [2007]	Mice	Pharyngeal aspiration	MWCNT—10, 20, or 40 µg	+	+ Including neuroinflammation of the brain	NR
Han et al. [2008b]	Mice	Pharyngeal aspiration with ozone exposure	MWCNT— 20 µg (eval: 5 and 24-hr post exp)	NR	+	NR
Hubbs et al. [2009]	Mice	Pharyngeal aspiration	MWCNT—20 or 80 µg (eval: 7 and 56 days post exp)	+	+	+
Sriram et al. [2009]	Mice	Pharyngeal aspiration	MWCNT—10 or 80 µg (eval: 1, 7, 28 days post exp)	NR	neuroinflammation	NR
Wolfarth et al. [2009]	Mice	Pharyngeal aspiration	MWCNT—40 µg (eval: 1, 7, 28, and 56 days post exp)	+	+	+
Porter et al. [2010]	Mice	Pharyngeal aspiration	MWCNT—10, 20, 40, or 80 µg (eval: 1, 7, and 28 days post exp)	+	+	+
Han et al. [2010]	Mice	Pharyngeal aspiration	MWCNT—20 or 40 µg (eval: 1 and 7 days post exp)	NR	+	NR
Mercer et al. [2011]	Mice	Pharyngeal aspiration	MWCNT—10, 20,40, or 80 µg (eval: 1, 7, 28, and 56 days)	+	+	+
Murray et al. [2012]	Mice	Pharyngeal aspiration	CNF—120 µg (eval: 1, 7, and 28 days post exp)	+	+	+

NR = Not Reported.
+ = effect observed.
– = no effect observed.

Table 3–7. Findings from published MWCNT short-term inhalation toxicology studies

Study design and exposure/dose				Observed pulmonary effects		
Author/ year	**Species**	**Exposure route**	**Exposure or dose**	**Granuloma**	**Inflammation**	**Fibrosis**
Li et al. [2007]	Mice	Inhalation	Est. lung deposition dose: 0.07, 0.14, .21 mg. (eval: at days 8, 16, and 24)	NR	–	NR
Mitchell et al. [2007]	Mice	Inhalation	0.3, 1, 5 $mg/m^3$6 hr/day for 7 or 14 days. (eval: at days 7 and 14)	–	–	–
Arkema [2008]	Rats	Head-nose inhalation	0.1, 0.5, 2.5 mg/m^3 6 hr/day for 5	– (0.1 mg/m^3) + (0.5, 2.5 mg/m^3)	– (0.1 mg/m^3) + (0.5, 2.5 $mg/m^{3)}$	–
Ryman-Rasmussen et al. [2009a]	Mice w/ preexisting Allergic inflammation	Nose-only inhalation	100 mg/m^3 for 6 hr (~10 mg/kg alveolar dose). (eval: at days 1 and 14)	Lung injury	+	+ when preexisting Allergic inflammation exists
Ma-Hock et al. [2009]	Rats	Head-nose inhalation	0.1, 0.5, 2.5 mg/m^3 6 hr/day–5 days/wk. for 13 weeks. (eval: at week 13)	+	+	–
Porter et al. [2009]	Mice	Whole body inhalation	10 mg/m^3 5 hr/day for 2, 4, and 8 days, then evaluated	+	+	+
Sriram et al. [2009]	Mice	Whole body inhalation	10 mg/m^3 5 hr/day for 2, 4, and 8 days, then evaluated	NR	Neuro-inflammation	NR
Ellinger-Ziegelbauer [2009]	Rats	Nose-only inhalation	11 and 241 mg/m^3 for 6 hr (eval: at days 7, 28, and 90)	NR	+	– (11 mg/m^3) + (241 mg/m^3)
Ryman-Rasmussen et al. [2009b]	Mice	Nose-only inhalation	1 or 30 mg/m^3 for 6 hr (~0.2 mg/kg and 4 mg/kg alveolar dose), (eval: at 1 day, and 2, 6, and 14)	–	– (1 mg/m^3) + (30 mg/m^3)	– (1 mg/m^3) + (30 mg/m^3)

Study design and exposure/dose				Observed pulmonary effects		
Author/ year	Species	Exposure route	Exposure or dose	Granuloma	Inflammation	Fibrosis
Pauluhn [2010a]	Rats	Nose-only inhalation	0.1, 0.4, 1.5 and 6 mg/m^3 for 6 hr, 5 days/week for 13 weeks	+(6 mg/m^3)	– (0.1 mg/m^3) + (0.4,1.5, 6 mg/m^3)	– (0.1 mg/m^3) (0.4 mg/m^3, focal septal thickening) +(1.5, 6 mg/m^3)

NR = not reported.
+ = effect observed.
– = no effect observed.

Table 3–8. Findings from published MWCNT or SWCNT short-term injection/implantation toxicology studies

Study design and exposure/dose				Observed pulmonary effects		
Author/ year	Species	Exposure route	Exposure or dose	Granuloma	Inflammation	Fibrosis
Deng et al. [2007]	Mice	Intravenous injection (also gavage)	1–600 µg MWCNT depending on exp. route	NR	–	NR
Takagi et al. [2008]	Mice	Intraperitoneal injection	27.5 % longer than 5 µm; 1×10^9 MWCNT/1 mL (corresponds to 3 mg) (eval: week 25)	mesothelioma	+	+
Poland et al. [2008]	Mice	Intraperitoneal injection	Long and short MWCNT 50 µg (eval: 1 and 7 days post exp)	Increase in response with increasing fiber length	Increase in response with increasing fiber length	NR
Muller et al. [2009]	Rats	Intraperitoneal injection	MWCNT < 1 µm on avg. length; 2 or 20 mg w/ defects, 20 mg wo/ defects. (eval: month 24)	No mesotheliomas		

Table 3–8. (Continued)

Study design and exposure/dose				Observed pulmonary effects		
Author/ year	**Species**	**Exposure route**	**Exposure or dose**	**Granuloma**	**Inflammation**	**Fibrosis**
Sakamoto et al. [2009]	Rats	Intrascrotal injection	MWCNT > 5 µm in length; 0.24 mg (1 mg/ kg body weight) 27.5 %	mesothelioma	NR	NR
Varga and Szendi [2010]	Rats	Peritoneal implantation (in gelatin capsule)	10 mg of MWCNT (1–2 µm length) or SWCNT (4–15 µm length)	No mesotheliomas	–	–
Liang et al. [2010]	Mice	Intraperitoneal injection	MWCNT (200 nm–2 µm length); 10, 50, or 250 mg/kg (eval: 28 days)	No mesotheliomas	- (10 and 50 mg/kg) + (250 mg/kg)	- (10 and 50 mg/kg) + (250 mg/ kg)
Murphy et al. [2011]	Mice	Intrapleural injection	MWCNT different lengths; 5 µg; (eval: 1 day, 1,4, 12 and 24 weeks)	No mesotheliomas	+ long MWCNT	+ long MWCNT
Nagai et al. [2011]	Rats	Intrapleural injection	MWCNT (mean lengths ~4-5 µm); 1 or 10 mg; (eval: up to 1 yr.) agglomerated and non-agglomerated	+ 1mg (non-agglomerated) mesothelioma at higher frequency than agglomerated	+ (1 and 10 mg)	+ (1 and 10 mg)

NR = not reported.

+ = effect observed.

– = no effect observed.

Respiratory effects after a subchronic inhalation exposure of rats to CNF (purity > 99.7%) were recently reported by DeLorme et al. [2012]. Both male and female Sprague Dawley rats were exposed nose-only inhalation to CNF (VGCFH Showa Denko), for 6 hrs/day, 5 days/week at concentrations of 0, 0.54, 2.5, or 25 mg/m^3 over a 90-day period and evaluated 1 day post exposure. Histopathological assessment included bronchoalveolar lavage fluid (BALF) analysis and cell proliferation studies of the terminal bronchiole, alveolar duct, and subpleural regions of the respiratory tract. The 25 mg/m^3 exposed rats and the non-exposed control group were also evaluated after a 3-month recovery period. The aerosol exposure to rats was characterized using SEM and TEM to determine the size distribution and fiber concentrations using NIOSH Method 7400. At an aerosol concentration of 0.54 mg/m^3 the fiber concentration was 4.9 fibers/cc with a MMAD of 1.9 μm (GSD 3.1), at 2.5 mg/m^3 the concentration was 56 fibers/cc with a MMAD of 3.2 μm (GSD 2.1), and at 25 mg/m^3 the concentration was 252 fibers/cc with a MMAD of 3.3 μm (GSD 2.0). The mean lengths and diameters of fibers were 5.8 μm and 158 nm, respectively with surface area measurements (by BET) of 13.8 m^2/g. At 1-day post exposure wet lung weights were significantly elevated compared to controls in male rats at 25 mg/m^3 and in female rats at 2.5 and 25 mg/m^3. Small increases in inflammation of the terminal bronchiole and alveolar duct regions were also observed in rats exposed to 2.5 mg/m^3 while histopathological assessments of rats exposed at 25 mg/m^3 found subacute to chronic inflammation of the terminal bronchiole and alveolar duct regions of the lungs along with thickening of the interstitial walls and hypertrophy/hyperplasia of type II pneumocytes. No adverse histopathological findings were reported for the 0.54 mg/m^3 exposure group. After the 3-month recovery period, lung weights remained elevated in each sex in the 25 mg/m^3 exposure group. Inflammation and the numbers (> 70%) of fiber-laden alveolar macrophages still persisted in the lung of rats exposed to 25 mg/m^3 with the inflammatory response reported to be relatively minor but significantly increased when compared to the non-exposed control group. Fibers were also observed to persist in the nasal turbinate's at 3-months post-exposure in all rats exposed at 25 mg/m^3 causing a nonspecific inflammatory response. In contrast to Murray et al. [2012], no fibrosis was noted in this inhalation study. The most likely reason for this discrepancy is a difference in alveolar lung burden between the Murray et al. [2012] and the DeLorme et al. [2012] study. In the former, the lung burden was 120 μg/mouse lung. In contrast, lung burden was not reported or estimated in the DeLorme et al. [2012] rat study. However, with a MMAD as large as 3.3 μm, nasal filtering would be expected to be high and alveolar deposition relatively low.

4. Conclusion—Hazard and Exposure Assessment

Results of laboratory animal studies with both SWCNT and MWCNT report qualitatively similar pulmonary responses including acute lung inflammation, epithelioid granulomas (microscopic nodules), and rapidly developing fibrotic responses at relatively low-mass doses (Section 3). Animal studies with CNT and CNF have shown the following:

1. Early onset and persistent pulmonary fibrosis in SWCNT-, MWCNT-, and CNF-exposed animals in shortterm and subchronic studies [Shvedova et al., 2005, 2008;

Mercer et al., 2008; Ma-Hock et al., 2009; Porter et al., 2010; Pauluhn 2010a; Mercer et al., 2011; Murray et al., 2012].

2. Similar pulmonary responses in animals (e.g., acute lung inflammation, interstitial fibrosis) when exposed to purified and unpurified SWCNT [Lam et al., 2004; Shvedova et al., 2005, 2008].
3. Equal or greater potency of SWCNT, MWCNT, and CNF compared with other inhaled particles (ultrafine carbon black, crystalline silica, and asbestos) in causing adverse lung effects including pulmonary inflammation and fibrosis [Shvedova et al., 2005; Muller et al., 2005; Murray et al., 2012].
4. CNT agglomeration affects the site of lung deposition and response; large agglomerates tend to deposit at the terminal bronchioles and proximal alveoli and induce a granulomatous response, while more dispersed structures deposit in the distal alveoli and cause interstitial fibrosis [Mercer et al., 2008; Porter et al., 2010]. Agglomerated SWCNT tend to induce granulomas, while more dispersed CNF and asbestos did not [Murray et al., 2012].
5. Intraperitoneal injection of long (> 5 μm) MWCNT in mice causes fibrotic lesions and mesothelial cell proliferation [Takagi et al., 2008; Murphy et al., 2011].

Although pulmonary responses to SWCNT and MWCNT are qualitatively similar, quantitative differences in pulmonary responses have been reported. In mice exposed to CNT by pharyngeal aspiration (10 μg/mouse), SWCNT caused a greater inflammatory response than MWCNT at 1 day post exposure [Shvedova et al., 2005, 2008; Porter et al., 2010]. Morphometric analyses indicate that well-dispersed purified SWCNT (< 0.23% iron) are not well recognized by alveolar macrophages (only 10% of the alveolar burden being within alveolar macrophages) [Shvedova et al., 2005], and that 90% of the dispersed SWCNT structures have been observed to cross alveolar epithelial cells and enter the interstitium [Mercer et al., 2008]. In contrast, approximately 70% of MWCNT in the respiratory airways are taken up by alveolar macrophages, 8% migrate into the alveolar septa, and 22% are found in granulomatous lesions [Mercer et al., 2010, 2011]. These findings suggest that well-dispersed SWCNT may be more potent in causing interstitial fibrosis on an equal mass lung burden basis than MWCNT [Mercer et al., 2008], possibly due to the greater tube count per mass of SWCNT [Mercer 2011; Wang et al., 2010a,b; Mercer et al., 2011]. In addition, although both SWCNT and MWCNT have been reported in the subpleural tissue of the lung [Mercer et al., 2008; Ryman-Rasmussen et al., 2009], penetration of the visceral pleura and translocation to the intrapleural space has been reported only for MWCNT [Hubbs et al., 2009; Mercer et al., 2010]. Despite these differences, CNTs of various types, both purified and unpurified, dispersed or agglomerated, all cause adverse lung effects in rats or mice at relatively low mass doses that are relevant to potential worker exposures.

Animal studies have also shown asbestos-type pathology associated with the longer, straighter CNT structures [Poland et al., 2008; Takagi et al., 2008; Murphy et al., 2011]. Mesothelial tumors have been reported in mice receiving intraperitoneal injection of long MWCNT (5–20 μm in length) [Takagi et al., 2008; Murphy et al., 2011]; whereas chronic bioassays of short MWCNT (avg. < 1 μm and < 2 μm in length, respectively) [Muller et al., 2009; Liang et al., 2010] did not produce mesothelioma. These findings are consistent with those reported by Yamashita et al. [2010] and Nagai et al. [2011] who found that MWCNT injected into the peritoneal cavity of mice or rats generated inflammation/genetic damage and mesothelioma that were related to the dimension of the CNT. Results from these peritoneal assay studies indicate that CNT of specific dimensions and durability can cause inflammation, fibrosis, and mesothelial tumors in mice and in rats; however, additional experimental animal research is needed to: (1) provide quantitative data on the biopersistence of different types of CNT in the lung and, (2) address the key question as to the precise dimensions (and possibly other physical-chemical characteristics) of CNT that pose a potential pathogenic risk for cancer including mesothelioma.

As synthesized, raw (unpurified) CNT, contain as much as 30% catalytic metals. Catalytic metals, such as iron-rich SWCNT, can generate hydroxyl radicals in the presence of hydrogen peroxide and organic (lipid) peroxides [Kagan et al., 2006], and when human epidermal keratinocytes cells are exposed to unpurified SWCNT (*in vitro* cellular studies), oxidant injury occurs [Shvedova et al., 2003]. These catalytic metals can be removed from raw CNT by acid treatment or by high temperature to yield purified CNT with low metal content. Removal of catalytic metals abolishes the ability of SWCNT or MWCNT to generate hydroxyl radicals. However, in laboratory animal studies the pulmonary bioactivity of SWCNT does not appear to be affected by the presence or absence of catalytic metals. Lam et al. [2004] compared the pulmonary response of mice to intratracheal instillation of raw (containing 25% metal catalyst) with purified (~2% iron) SWCNT and found that the granulomatous reaction was not dependent on metal contamination. Likewise, the acute inflammatory reaction of mice after aspiration of raw (30% iron) versus purified (< 1% iron) SWCNT was not affected by metal content [Shvedova et al., 2005, 2008].

Pulmonary exposure to CNT have shown systemic responses including an increase in inflammatory mediators in the blood, as well as oxidant stress in aortic tissue and increase plaque formation in an atherosclerotic mouse model [Li et al., 2007; Erdely et al., 2009]. Pulmonary exposure to MWCNT also depresses the ability of coronary arterioles to respond to dilators [Stapleton et al., 2011]. These cardiovascular effects may be due to neurogenic signals from sensory irritant receptors in the lung. Mechanisms, such as inflammatory signals or neurogenic pathways causing these systemic responses, are under investigation.

Results from *in vitro* cellular studies have shown that SWCNT can cause genotoxicity and abnormal chromosome number, because of interference with mitosis (cell division), by disrupting the mitotic spindles in dividing cells and inducing the formation of anaphase bridges among the nuclei [Sargent et al., 2009]. In vitro studies also indicate that exposure to CNF can cause genotoxicity (micronuclei) as a result of reactive oxygen species (ROS) production, which in turn reacts with DNA, and by interfering physically with the DNA/chromosomes and/or mitotic apparatus [Kisin et al., 2011]. Low-dose, long-term exposure of bronchial epithelial cells to MWCNT has been shown to induce cell transformation, and these transformed cells induce tumors after injection into nude mice [Stueckle et al., 2011; Wang et al., 2011].

Currently, there are no studies reported in the literature on the adverse health effects in workers producing or using CNT or CNF. However, because humans can also develop lung inflammation and fibrosis in response to inhaled particles and fibers, it is reasonable to assume that at equivalent exposures (e.g., lung burden/alveolar epithelial cell surface) to CNT and CNF, workers may also be at risk of developing these adverse lung effects.

Although data on workplace exposures to CNT and CNF are limited, aerosolization of CNT and CNF has been shown to occur at a number of operations during research, production, and use of CNT and CNF, including such work tasks as transferring, weighing, blending, and mixing. Worker exposure to airborne CNT and CNF has frequently been observed to be task-specific and short-term in duration, with exposure concentrations (frequently reported as particle number or mass concentrations) found to exceed background exposure measurements when appropriate engineering controls are not used to reduce exposures [Maynard et al., 2004; Methner et al., 2007; Han et al., 2008a; Bello et al., 2009; Tsai et al., 2009; Bello et al., 2010; Evans et al., 2010; Johnson et al., 2010; Lee et al., 2010; Cena and Peters 2011; Dahm et al., 2011]. Results from studies also suggest that the airborne concentration and the physical-chemical characteristics of particles (e.g., discrete versus agglomerated CNT) released while handling CNT may vary significantly with production batch and work process. Comprehensive workplace exposure evaluations are needed to characterize and quantify worker exposure to CNT and CNF at various job tasks and operations, and to determine what control measures are the most effective in reducing worker exposures.

The findings of adverse respiratory effects (i.e., pulmonary fibrosis, granulomatous inflammation) and systemic responses in animals indicate the need for protective measures to reduce the health risk to workers exposed to CNT and CNF. Available evidence also indicates that the migration of MWCNT into the intrapleural space could potentially initiate mesothelial injury and inflammation that over time cause pleural pathology, including mesothelioma. Long-term inhalation studies are needed to determine whether CNT and CNF of specific dimension and chemistry can cause cancer in laboratory animals at doses equivalent to potential workplace exposures. In addition, the potential for migration of CNT through the lungs and for accumulation in the intrapleural space with time after inhalation requires further investigation. Until results from animal research studies can fully explain the mechanisms in which inhalation exposure to CNT and CNF cause adverse lung effects and possible systemic effects, all types of CNT and CNF should be considered an occupational respiratory hazard, and the following actions should be taken to minimize health concerns:

1. Minimize workplace exposures.
2. Establish an occupational health surveillance program for workers exposed to CNT and CNF (Section 6, Appendix B).

5. CNT Risk Assessment and Recommended Exposure Limit

5.1. Risk Assessment and Recommended Exposure Limit (REL)

NIOSH bases its recommended exposure limits (RELs) on quantitative risk assessments when possible. Quantitative risk assessment provides estimates of the severity and likelihood of an adverse response associated with exposure to a hazardous substance. The hazard and quantitative risk assessments (Section 4 and Appendix A) provide the health basis for developing a recommended exposure limit (REL) for CNT and CNF. Establishing healthbased exposure limits is the first consideration by NIOSH in setting a REL. The analytical feasibility of measuring worker exposures to airborne CNT and CNF is also taken into account in the establishment of the REL (Section 6.1).

In general, quantitative risk assessment involves the following steps: first a data set is selected that best depicts a dose-response relationship, in this case, the relationship between exposure to CNT and pulmonary effects in animals. Then, a critical dose in the animal lungs is calculated. A frequently used indicator of critical dose is the benchmark dose (BMD) which is defined as the dose corresponding to a small increase in response (e.g. 10%) over the background level of response [Crump 1984]. Next, the dose in humans, that is equivalent to the critical dose in the animals, is estimated. This requires adjusting for species differences between animals and humans. It is assumed in the absence of specific data that an equivalent dose in animals and humans will result in the same risk of disease, based on the assumption that the same mechanism of action is operating in both animals and humans. After the critical average dose in human lungs is estimated from the animal data, an equivalent workplace concentration over a full working life-time is derived. This is accomplished by using mathematical and physiological models to estimate the fraction of the dose that reaches various parts of the respiratory tract and is deposited and cleared [Kuempel et al., 2006; Schulte et al., 2010; NIOSH 2011a]. Variability in human dose and response, including sensitive subpopulations, and uncertainty in the extrapolating animal data to humans are typically addressed with uncertainty factors in the absence of specific data.

NIOSH determined that the best data to use for a quantitative risk assessment and as the basis for a REL were the nonmalignant pulmonary data from shortterm and subchonic animal studies. In these studies, lung exposures to CNT (i.e., various types of MWCNT and SWCNT, purified and unpurified, dispersed or agglomerated, and with different metal content) were observed to cause earlystage adverse lung effects including, pulmonary inflammation, granuloma, alveolar septal thickening, and pulmonary fibrosis (Section 3 and Appendix A). NIOSH considers these animal lung effects to be relevant to workers because similar lung effects have also been observed in workers with occupational lung disease associated with exposure to various types of inhaled particles and fibers [Rom and Markowitz 2006; Hubbs et al., 2011]. Human-equivalent risk estimates were derived from animal dose-response data (in rats and mice). Human-equivalent exposures over a 45-year working lifetime were estimated to be associated with either a specified risk level (e.g., 10%) of earlystage lung effects or with a no observed adverse effect level based on the animal studies. In the absence of validated lung dosimetry models for CNT, lung doses were estimated using spherical particle-based models and CNT airborne size data, assuming either deposited or retained lung dose in animals or humans.

The findings from this analysis indicate that workers are potentially at risk of developing adverse lung effects if exposed to airborne CNT during a 45-year working lifetime. Table 5–1 provides a summary of the estimated exposure concentrations (8-hour TWA) associated with 10% excess risk based on the animal data. Table 5–2 provides a summary of the risk estimates at the REL of 1 $\mu g/m^3$ (8-hour TWA).

Working lifetime exposures to 0.2–2 $\mu g/m^3$ (8-hour TWA concentration) were estimated to be associated with 10% excess risk of early stage lung effects (minimal granulomatous inflammation or alveolar septal thickening, grade 1 or higher) (95% lower confidence limit, LCL estimates) based on results from the subchronic animal inhalation studies with MWCNT [Ma-Hock et al., 2009; Pauluhn 2010a] (Tables 5–1 and A–5). For slight/mild (grade 2 or higher) lung effects, the working lifetime exposure estimates are 0.7–19 $\mu g/m^3$ (8-hour T WA concentration; 95% LCL estimates) (Tables 5–1 and A–6). Risk estimates derived from other animal studies (e.g., single dose with up to 90-day followup) using SWCNT and other types of MWCNT (Tables A–3 and A–4) are consistent with these estimates, i.e., 0.08–12 μg/m^3 (8-hour TWA) (95% LCL estimates). These working lifetime exposure concentration estimates vary by approximately two orders of magnitude (across the different types of CNT, study design, animal species/strain and gender, route of exposure, and response endpoints); yet all of these estimates are relatively low airborne mass concentrations, most within ~1–10 μg/m^3 (8-hour TWA). NIOSH does not consider a 10% estimated excess risk over a working lifetime to be acceptable for these earlystage lung effects, and the REL is set at the optimal limit of quantification (LOQ) of the analytical method carbon (NIOSH method 5040) (Appendix C). Additional estimates were derived from the no observed adverse effect level (NOAEL) or lowest observed adverse effect level (LOAEL) in the rat subchronic inhalation studies of MWCNT [Pauluhn 2010a; Ma-Hock et al., 2009] (Section A.6.2 and A.6.3). The human-equivalent working lifetime concentrations of ~4-18 μg/m^3 (8-hour TWA) were estimated to be equivalent to the rat subchronic NOAEL or LOAEL of 0.1 mg/m^3 (Table A–13). Applying data-appropriate uncertainty factors (e.g., Table A–14) to estimate safe working lifetime exposure (essentially zero risk assuming a threshold model) results in concentrations of less than 1 μg/m^3 (8-hour TWA) over a 45-yr working lifetime.

Table 5–1. Estimated exposure concentration associated with a 10% risk of adverse lung effects above background*

Lung disease indicator†	Estimated working lifetime exposure concentration (8-hr TWA)‡	
	Maximum likelihood estimate	**95% Lower confidence limit estimate**
Minimal lung effects (grade 1 or higher)	0.5 to 4 μg/m^3	0.2 to 2 μg/m^3
Slight or mild lung effects (grade 2 or higher)	1 to 44 μg/m^3	0.7 to 19 μg/m^3

Abbreviation: TWA=Time-weighted average.

* Excess (exposure-attributable) risk during a 45-year working lifetime.

† Histopathology findings of granulomatous inflammation [Ma-Hock et al., 2009] or alveolar septal thickening [Pauluhn 2010] in rat subchronic inhalation studies of multiwall carbon nanotubes.

‡ Estimates vary by rat study and lung burden estimation method (Appendix A, Tables A–7 and A–8).

Table 5–2. Estimated risk of adverse lung effects at recommended exposure limit of 1 μg/m^3 (8-hour TWA) during a 45-year working lifetime

Lung disease indicator*	Excess risk†	
	Maximum likelihood estimate	**95% Lower confidence limit estimate**
Minimal lung effects (grade 1 or higher)	2.4% to 33%	5.3% to 54%
Slight or mild lung effects (grade 2 or higher)	0.23% to 10%	0.53% to 16%

Abbreviation: TWA=Time-weighted average.

* Histopathology findings of granulomatous inflammation [Ma-Hock et al., 2009] or alveolar septal thickening [Pauluhn 2010] in rat subchronic inhalation studies of multiwall carbon nanotubes.

† Exposure-attributable risk (added risk above background). Estimates vary by rat study and lung burden estimation method (Appendix A, Tables A–7 and A–8).

A more detailed summary of the working lifetime exposure estimates—shown by animal study, lung response, and effect level estimate—is provided in Table 5–3. Table 5–4 summarizes the factors, assumptions, and options involved in the CNT risk assessment (see Appendix A for complete information). Evaluations of the variability and uncertainty in these risk estimates are provided in Sections 5.3, A.4, and A.6.

Among the uncertainties in this risk assessment using animal data, there is uncertainty in extrapolating the respiratory effects observed in shortterm or subchronic animal studies to estimate the probability of chronic respiratory effects in humans. In the absence of chronic data, these animal studies provide the best available information to derive initial estimates of health risk for use in REL development. Subchronic (13 wk.) exposure studies are a standard toxicity assay used in human health risk assessment, although the studies with shorter exposure and post-exposure durations also provide useful information about the relationship between CNT lung dose and response. To the extent that the precursor effects to chronic disease are observed in these shorter-term studies, the hazard and risk estimates would be expected to provide useful information for chronic disease prediction and prevention. Although there is uncertainty in the benchmark dose estimates from the subchronic studies because of the dose-spacing and high response proportions, these estimates are similar to the NOAEL and LOAEL values reported in these studies [Ma-Hock et al., 2009; Pauluhn 2010a] (Table A–12).

One of the measures of pulmonary fibrosis used in the shorter-term studies [Shvedova et al., 2005, 2008; Mercer et al., 2008, 2011]—alveolar epithelial cell thickness (due to collagen deposition)—was previously used in the U.S. EPA ozone standard. This biological response was selected by EPA as the adverse lung response for crossspecies dose-response extrapolation because it indicates "fundamental structural remodeling" [US EPA 1996; Stockstill et al., 1995].

Some of these studies provide data comparing the potency of CNT with that of other particles or fibers for which animal and human data are available on the long-term adverse health effects. These studies show that on a mass basis, CNT had equal or greater potency (pulmonary inflammation or fibrosis response at a given mass dose) to that of ultrafine carbon black, crystalline silica, or chrysotile asbestos [Lam et al., 2004; Muller et al., 2005; Shvedova et al., 2005]. These comparative toxicity findings between CNT and other well-studied particles or fibers help to reduce the uncertainty about whether the lung effects in these short-term studies are relevant to evaluating the chronic respiratory hazard of CNT.

Based on currently available data, it is difficult to assess the relative toxicity of the various types of CNT and CNF because there has been limited systematic study of various CNT and CNF using the same study design. These available studies differ in factors that include the rodent species and strain, the techniques and assays for measuring lung effects, and the exposure and post-exposure durations. Despite differences in the type and composition of SWCNT and MWCNT used in the animal studies, the risk estimates across the different types of CNT and studies are associated with relatively low mass exposure concentrations. Although data from laboratory animal studies with CNF are limited, the similarities in physical-chemical properties and adverse lung effects between CNF and CNT support the need to control exposures to CNF at the REL derived for CNT.

Table 5–3. Summary of human-equivalent working lifetime exposure concentration estimates associated with animal effect levels

Animal effect level	Animal lung response	Working lifetime 8-hr TWA (μg/m³)		Animal study	Table in CIB
		Assuming estimated deposited lung dose*	Assuming estimated retained lung dose*		
Subchronic studies					
NOAEL	No statistically significant lung response	nd	3.5	Pauluhn 2010a	A–13
LOAEL	Minimal granulomatous inflammation (grade 1+)	nd	4.0	Ma-Hock et al., 2009	A–13
BMD_{10}	Minimal granulomatous inflammation or alveolar	0.51, 0.77	2.7, 4.2	Ma-Hock et al., 2009; Pauluhn	A–5
$BMDL_{10}$	septal thickening (grade 1+)	0.19, 0.38	1.0, 1.9	2010a, respectively	
BMD_{10}	Slight/mild granulomatous inflammation or alveolar	1.0, 6.4	6.2, 44	Ma-Hock et al., 2009; Pauluhn	A–6
$BMDL_{10}$	septal thickening (grade 2+)	0.69, 3.3	4, 19	2010a, respectively	
Short-term studies					
BMD_{10}	Alveolar connective tissue thickness	0.11, 1.8, 4.7	nd	Shvedova et al., 2005; Mercer et al., 2011; Shvedova et al., 2008, respectively	A–3
$BMDL_{10}$		0.075, 1.0, 2.5			
BMD_{10}	Hydroxyproline amount	18	nd	Muller et al., 2005	A–3
$BMDL_{10}$		12			
BMD_{10}	Granuloma	10	nd	Lam et al., 2004	A–4
$BMDL_{10}$		1.7			

Abbreviations: nd=not determined; BMD_{10}=benchmark dose (maximum likelihood estimate) associated with 10% additional risk of the adverse response; $BMDL_{10}$=95% lower confidence limit estimate of the BMD_{10}; LOAEL= lowest observed adverse effect level; NOAEL=No observed adverse effect level

* Interspecies dose normalized by alveolar surface area in animal and humans (Section A.2.3.4).

Table 5–4. Factors, assumptions, and options evaluated in the CNT risk assessment

Factor	Assumption	Options evaluated
CNT type	Risk estimates for various types of CNT are relevant to worker exposures, based on animal dose-response data of administered or estimated lung dose and early-stage, persistent lung responses, using standardized methods	SWCNT (Fe 2%) SWCNT (Fe 0.2–0.3%) SWCNT (Fe 18%) MWCNT (Co 2%, Fe 0.5%) MWCNT (Al_2O_3 9.6%) MWCNT (Co 0.5%)
Route of exposure	Lung dose is associated with animal response regardless of route of exposure and is relevant to human inhaled dose	Inhalation Pharyngeal aspiration Intratracheal instillation
Duration of exposure	Subchronic or short-term exposure is associated with observed lung responses and relevant to humans	13-week inhalation, 1d–26wk PE Short-term inhalation, 56d PE Single dose, 28–91d PE
Species/strain	Animal model is relevant for humans	Rat • Sprague-Dawley • Wistar • Mouse • B6C3F1 • C57BL/6
Sex	No sex-specific effect	Male Female
Critical effect	Animal response is relevant to humans	Dichotomous response: Granuloma Granulomatous inflammation* • grade 1+ • grade 2+ Lipoproteinosis • grade 1+ Alveolar septal thickening • grade 1+ • grade 2 Continuous response: Alveolar connective tissue (septal) thickness Hydroxyproline amount
Critical effect level	Dose to target tissue (as administered, estimated deposited, or estimated retained mass in alveolar region of lungs) is associated with lung response	BMDL BMD NOAEL LOAEL
Human equivalent dose	Humans would have an equal lung response (on average) to an estimated equivalent alveolar dose	Alveolar surface area
• Interspecies normalization		Alveolar macrophage cell volume
• Duration of exposure	Working lifetime average cumulative exposure would be associated with the human-equivalent lung response	45-years (8-hr TWA, 40-hr workweek, 50 wk/yr)

Table 5–4. (Continued)

Factor	Assumption	Options evaluated
• Clearance kinetics	True average working lifetime exposure lies between deposited and retained CNT dose estimates based on spherical particle dosimetry model	Deposited dose (no clearance)
		Retained dose (normal clearance)
• Ventilation rate	Workers on average breath the same amount of air in a day; normal oronasal augmenter breathing pattern	Reference worker (9.6 m^3/d)
• Alveolar deposition fraction	Airborne particle size distribution predicts deposition fraction	Spherical particle-based model (MPPD), incl. two versions • Density of 1 • Density of <1

* Grade 1: minimal; grade 2: slight/mild.

NIOSH is recommending an occupational exposure limit for CNT and CNF to minimize the risk of developing adverse lung effects over a working lifetime. A mass-based airborne exposure limit is being recommended because this exposure metric is the same as that used in determining the dose-response relationship in animal studies and deriving the risk estimates, as well as being the most common exposure metric currently used in monitoring workplace exposures to CNT and CNF. The REL is based on the respirable particle-size fraction because the adverse lung effects in the animal studies were observed in the alveolar (gas-exchange) region. "Respirable" is defined as the aerodynamic size of particles that, when inhaled, are capable of depositing in the alveolar region of the lungs [ICRP 1994]. Sampling methods have been developed to estimate the airborne mass concentration of respirable particles [ACGIH 1984; CEN 1993; ISO 1995; NIOSH 1998]. Although the goal is to establish a REL that would eliminate any potential risk for developing respiratory disease, limitations exist in reliably measuring airborne CNT and CNF using currently available mass-based sampling and analytical methods. NIOSH is recommending that NIOSH Method 5040 [NIOSH 1994; Birch 2004a, b] be used to measure workplace airborne exposure to respirable CNT and CNF.

An upper estimate of the LOQ of Method 5040 (7 μg/m^3) was proposed as the draft REL (8-hr TWA) for CNT and CNF [NIOSH 2010]. This upper limit was based on total carbon (TC) results found for filter media from different vendors and lots, by different laboratories, and during a 6-month period. In practice, when elemental carbon (EC) results are used from media blanks submitted with the samples to estimate the LOQ, a much lower value can be achieved (Section 6.1). It is important to note that the LOQ for NIOSH Method 5040 depends on the media blank variability, filter area sampled, portion of filter analyzed, and the collected sample air volume. As discussed in Section 6.1, under optimum conditions an LOQ of 1 μg/m^3 can be obtained for an 8-hr respirable sample collected on a 25-mm filter at a flow rate of 4 liters per minute (lpm).

Considering the potential uncertainties in estimating the health risks and current limitations of analytical methodologies, NIOSH is recommending a REL of 1 μg/m^3 EC as an 8-hr TWA airborne respirable mass concentration for up to a 40-hr work week. For working lifetime exposures at 1 μg/m^3, the MLE risk estimates for slight/mild level of lung effects (based on the rat subchronic inhalation studies of MWCNT) range from 0.23% to 10%, and the 95% UCL estimates range from 0.53% to 16% depending on the rat study and the assumptions used in estimating the CNT lung dose (Tables 5–2 and A–8). The more sensitive endpoint of minimal level of lung effects in the rat subchronic inhalation studies were associated with MLE risk estimates of 2.4% to 33% and 95% UCL estimates of 5.3% to 54% (Tables 5–2 and A–7). Estimates of a 45-yr working life-time no-effect concentration (8-hr TWA) based on the NOAEL or LOAEL estimates from the rat subchronic studies are < 1 $\mu g/m^3$ (8-hr TWA) (Section A.6.3). The estimates based on the shortterm studies of SWCNT and MWCNT in rats and mice are consistent with those from the subchronic studies of MWCNT in rats (Section A.3).

NIOSH recognizes that the REL may not be completely health protective and that there is uncertainty in these risk estimates, but maintaining exposures below the REL should help to lower workers' risk of developing occupational lung disease over a working lifetime. The REL and other recommendations in this CIB should also assist employers in establishing an occupational health surveillance program that includes elements of hazard and medical surveillance. Until improved methods to measure airborne exposures to CNT and CNF are established, continued efforts should be made to reduce airborne concentrations to as low as possible below the REL. Approaches to optimize the sampling and analysis of exposures are discussed in Section 6.1 and Appendix C. Examples of engineering controls to reduce or eliminate workers' exposures to CNT and CNF are provided in Section 6.2. Additional guidance, including personal protective equipment, respirators, training, and medical surveillance is provided in Sections 6.3 through 6.7. Based on available workplace exposure data, it is not possible for NIOSH to determine whether the NIOSH REL can be achieved in all workplaces where exposure to CNT and CNF occur; however, exposure data that have been reported indicate that implementing appropriate engineering control measures (e.g., local exhaust ventilation, enclosures) can eliminate or greatly reduce worker exposures [Han et al., 2008a; Methner et al., 2008; Tsai et al., 2009; Johnson et al., 2010; Lee et al., 2010; Cena and Peters 2011; Dahm et al., 2011].

5.2. Other Derived Occupational Exposure Limits for CNT

One of the earliest OELs for CNT was proposed by the British Standards Institute [BSI 2007]—the benchmark exposure limit (BEL) of 0.01 fiber/cm^3, or one-tenth of their asbestos exposure limit (Table 5–5). Nanocyl [2009] derived an estimated OEL of 2.5 $\mu g/m^3$ for an 8-hr TWA exposure based on applying an overall assessment (a.k.a. uncertainty) factor of 40 to the LOAEL of 0.1 mg/m^3 in the Ma-Hock et al. [2009] subchronic rat inhalation study of MWCNT. Aschberger et al. [2010] proposed OELs of 1 $\mu g/m^3$ for MWCNT studied by Ma-Hock et al. [2009] and 2 $\mu g/m^3$ for MWCNT from Pauluhn [2010a], by adjusting 0.1 mg/m^3 (the LOAEL in Ma-Hock et al. [2009] and the NOAEL in Pauluhn [2010a]) for rat-to-human daily exposure and respiratory volume, and applying an overall assessment factor of 50 and 25, respectively.

Pauluhn [2010b] derived an OEL using subchronic data in rats inhaling MWCNTs (Baytubes®) [Pauluhn 2010a]. This approach was based on the biological mechanism of volumetric overloading of alveolar macrophage-mediated clearance of particles from the lungs of rats [Morrow 1988]. Increased particle retention half-time (an indication of lung clearance overload) was reported in rats exposed by subchronic inhalation to MWCNT

(Baytubes®) at 0.1, 0.4, 2.5, or 6 mg/m^3 The overloading of rat lung clearance was observed at lower-mass doses of MWCNT (Baytubes®) compared with other poorly soluble particles; and the particle volume dose was better correlated with retention half-time among poorly soluble particles including CNT [Pauluhn 2010a, b]. Pauluhn [2010b] reported benchmark concentration (BMC) estimates of 0.16 to 0.78 mg/ m^3 for rat lung responses of pulmonary inflammation and increased collagen, but selected the lower NOAEL of 0.1 mg/m^3 to derive a human-equivalent concentration. The NOAEL was adjusted for human and rat differences in factors affecting the estimated particle lung dose (i.e., ventilation rate, alveolar deposition fraction, retention kinetics, and total alveolar macrophage cell volume in each species). The product of these ratios resulted in a final factor of 2, by which the rat NOAEL was divided, to arrive at a human-equivalent concentration of 0.05 mg/m^3 (8-hr TWA) as the OEL for MWCNT (Baytubes®). No uncertainty factors were used in deriving that estimate.

Table 5–5. Recommended occupational exposure limits for CNT

Reference	Occupational exposure limit (OEL)	Comments
Pauluhn [2010b]	0.05 mg/m^3 (8-hr TWA) for MWCNT (Baytubes®)	Based on rat subchronic (13-wk) inhalation study of MWCNT (Baytubes®) and prevention of lung clearance overload and associated pulmonary effects. Rat NOAEL of 0.1 mg/m^3 adjusted by a factor of 2 for worker exposure day, air intake, deposition, and clearance kinetics. No uncertainty factors were applied.
Nakanishi (ed) [2011a,b]	30 μg/m^3 (8-hr TWA) for CNT	Based on 4-wk inhalation studies of SWCNT and MWCNT in rats. Lowest NOAEL of 0.13 mg/m^3 (for high surface area SWCNT) used as basis for CNT OEL). Adjusted for worker exposure day, air intake, deposition fraction, and body weight; uncertainty factor of 6. OEL is period-limited (15-yr).
Nanocyl [2009]	2.5 μg/m^3 (8-hr TWA) for MWCNT	Adjusted rat LOAEL of 0.1 mg/m^3 (subchronic inhalation) [Ma-Hock et al. 2009] to workers and applied assessment factor of 40.
Aschberger et al. [2010]	2 μg/m^3 (8-hr TWA) for MWCNT	Adjusted rat NOAEL of 0.1 mg/m^3 (subchronic inhalation) [Pauluhn 2010a] for worker exposure day and air intake; assessment factor of 25.
	1 μg/m^3 (8-hr TWA) for SWCNT	Adjusted rat LOAEL of 0.1 mg/m^3 (subchronic inhalation) [Ma-Hock et al. 2009] for worker exposure day and air intake; assessment factor of 50.
BSI [2007]	0.01 fibers/ml for fibrous nanomaterials with high aspect ratios (> 3:1 and length > 5000 nm)	Benchmark exposure level (BEL) based on one tenth of the asbestos exposure limit

The Japanese National Institute of Advance Industrial Science and Technology (AIST) derived an OEL for CNT of 30 μg/m^3 [Nakanishi 2011a,b], based on studies supported by the New Energy and Industrial Technology Development Organization (NEDO) of Japan. Rat NOAELs for pulmonary inflammation were identified in 4-week inhalation studies of SWCNT and MWCNT [Morimoto et al. 2011a,b]. Human-equivalent NOAELs were estimated by accounting for rat and human differences in exposure duration, ventilation rate, particle deposition fraction, and body weight [Nakanishi 2011b]. The rat NOAELs of 0.13 and 0.37 mg/m^3 for SWCNT and MWCNT, respectively, were estimated to be equivalent to 0.03 and 0.08 mg/m^3 in humans including adjustment by an uncertainty factor of 6. This total uncertainty factor included a factor of 2 for uncertainty in subchronic-to-chronic extrapolation and a factor of 3 for uncertainty in rat to human toxicokinetic differences (factors of 1 were assumed for toxicodynamic differences in rats and humans and for worker inter-individual variability). A relationship was reported between the BET specific surface area of various types of CNT and pulmonary inflammation (percent neutrophils in bronchoalveolar lavage fluid) (Figure V.2 in Nakanishi [2011b]). Thus, the OEL of 0.03 mg/m^3 was proposed for all types of CNT, based on the data for the SWCNT with the relatively high specific surface area of ~1,000 m^2/g (which was noted would be more protective for other CNTs with lower specific surface area). A period-limited (15-yr) OEL was proposed due to uncertainty in chronic effects and based on the premise that the results will be reviewed again within that timeframe with further data [Nakanishi 2011a].

In summary, these currently proposed OELs for CNT range from 1 to 50 μg/m^3 (8-hr TWA concentration) [Aschberger et al. 2010; Nanocyl 2009; Pauluhn 2010b; Nakanishi (ed) 2009a], including the NIOSH REL of 1 μg/m^3. Despite the differences in risk assessment methods and assumptions, all of the derived OELs for CNT are low airborne mass concentrations relative to OELs for larger respirable carbon-based particles. For example, the current U.S. OELs for graphite or carbon black are approximately 2.5 to 5 mg/m^3. Each of these CNT risk assessments supports the need to control exposures to CNT in the workplace to low airborne mass concentrations (μg/m^3) to protect workers' health.

5.3. Evaluation of Uncertainties in CNT Risk Assessment and REL

Animal data have been used to evaluate the health hazard and risk of occupational exposure to CNT and CNF. Limited human exposure data are available, and NIOSH is not aware of any studies or reports at this time of any adverse health effects in workers producing or using CNT or CNF. The best available scientific information to develop recommended exposure limits is from the subchronic (13-wk) animal inhalation studies of two types of MWCNT and the shorter-term animal studies of SWCNT and other types of MWCNT.

The analysis of animal data in this risk assessment includes: (1) identifying the adverse health effects that are associated with exposure to CNT or CNF in laboratory animals; (2) evaluating the severity of the response and the relevance to humans; and (3) estimating the human-equivalent dose and likelihood (risk) of adverse effects in workers. Ideally, sufficient evidence is desired to derive exposure limits that are estimated to be associated with essentially a zero risk of an adverse health effect even if exposed 8 hr/d, 40 hr/wk. over a 45-yr working lifetime. However, limitations in the scientific data result in uncertainties about those hazards and risk estimates. Characterizing the degree of that uncertainty, and the extent to which use of those data are useful for occupational health risk management decision-making, is an important step in risk assessment. Alternative models and methods contribute to the differences in the risk estimates, and there is uncertainty about which biological endpoints, animal models, and interspecies and dose rate extrapolation methods may be most predictive of possible human health outcomes.

5.3.1. Strength of Evidence for Estimating a Health-Based REL for CNT and CNF

NIOSH and others have used the published animal dose-response data to develop OELs for various types of CNT (see Section 5.2). The methods and assumptions differ across these studies and have resulted in mass-based OELs ranging from 1 µg/m^3 to 50 µg/m^3. A proposed benchmark exposure limit (BEL) of 0.01 fiber/cm^3 for CNT has also been proposed [BSI 2007].

In the NIOSH risk assessment (Appendix A), the animal effect level estimates (and also the equivalent working lifetime estimates) for early-stage noncancer lung effects (granulomatous inflammation, interstitial thickening, or fibrosis) differ by approximately two orders of magnitude depending on the animal study, CNT type, and lung dose estimation methods (Tables A–3 through A–6; Table A–13). However, these estimated human-equivalent 45-yr working lifetime exposure concentrations (8-hr TWA) are all relatively low airborne mass concentrations (approximately 0.1–19 µg/m^3 before adjusting for uncertainty in these estimates).

The major areas of uncertainty in the CNT risk assessment include: (1) the critical effect or lung response measure including level of severity; (2) dose rate and retention assumptions for extrapolation from subchronic or short-term animal studies to chronic exposure in humans; and (3) low dose extrapolation using the benchmark dose models to estimate risks below the 10% benchmark dose. The relatively minor areas of uncertainty include: (1) benchmark dose estimation; (2) impact of route of exposure; and (3) rat lung dose estimation.

These uncertainties can result in either under-estimation or over-estimation of the true health risk to workers at a given exposure scenario. Each of these areas is discussed further below.

5.3.2. Major Areas of Uncertainty

(1) Lung response and severity level

The REL is based on estimates of excess risk of earlystage noncancer lung effects, which NIOSH has determined are relevant to human health risk assessment (Section A.2.1.3). The extent to which these lung responses would be associated with functional deficits in animals or clinically significant effects in humans is uncertain. However, these lung responses include early onset fibrosis which persisted or progressed after the end of exposure [Shvedova et al. 2005. 2008; Porter et al. 2010; Mercer et al. 2011]. Limited evidence in animals suggests that these effects may be associated with some lung function decrement (reduced breathing rate in mice) [Shvedova et al. 2008]. A quantitative measure of pulmonary fibrosis—alveolar interstitial (septal, or connective tissue) thickening [Shvedova et al. 2005, 2008; Mercer et al. 2011]—was previously used in developing the health basis for the U.S. air contaminant standards for another lung toxicant, ozone [US EPA 1994]. EPA selected this health endpoint as the critical lung effect in animals and extrapolated to humans the lung dose associated with this effect [US EPA 1996; Stockstill et al. 1995]. Additional measures of early stage lung effects in rats and mice that were used in this CNT risk assessment include minimal or greater alveolar septal thickening, granulomas, or granulomatous inflammation [Pauluhn 2010a; Ma-Hock et al. 2009; Lam et al. 2004].

The choice of health endpoint and severity level from the subchronic studies resulted in different REL estimates by factors of several-fold to an order of magnitude (Table A–5 and A–6). In addition to quantitative differences in risk estimates for non-cancer effects, there is also qualitative uncertainty about the risk of other disease endpoints, including cancer. Possible cancer risk (e.g., associated with fiber-like structures) is an area of considerable uncertainty for CNT and CNF which warrants targeted research and a high level of exposure control until the risk is understood [Schulte et al. 2012].

(2) Dose rate and retention

Appendix A–6 provides some analyses to show the quantitative influence of dose rate and lung retention assumptions on the risk estimates and REL derivation (Tables A–5 and A–6; Section A.6.3.2.2). The lung effects were assumed to be associated with the total lung dose, regardless of the dose rate. If the average daily deposited lung dose is assumed (i.e., no difference in rat or human clearance rates), then the human-equivalent concentration would be ~30 times higher than that based on the ICRP [1994] clearance model, and ~10 times higher than that assuming simple first-order kinetics [Snipes et al. 1989; Pauluhn 2010b]. The human-equivalent working lifetime (8-hr TWA) estimates based on deposited lung dose (assuming no clearance) are lower by a factor of ~5–7 than those estimates based on retained lung burden (assuming normal clearance) (Tables A–5 and A–6).

(3) Inter-species dose normalization

Alternative assumptions about the biologically-relevant measure of equivalent dose can result in considerable differences in the human-equivalent dose. NIOSH normalized the interspecies lung dose based on the ratio of the human-to-animal average alveolar surface area. Alternatively, Pauluhn [2010b] normalized the dose from rat to humans based on the average total alveolar macrophage cell volume. This difference resulted in a factor of ~4 in the human-equivalent lung burdens and working lifetime 8-hr TWA concentration estimates (Table A–13).

(4) Low dose extrapolation

All of these animal data and methods result in low equivalent working lifetime exposure estimates. The animal NOAEL, LOAEL, and BMD(L)* estimates, and the equivalent human working lifetime exposure estimates, indicate low mass concentrations (8-hr TWA) over a 45-year working lifetime (<0.1—19 μg/m^3) (95% LCL, Tables A–3 through A–6). As discussed in Section A.2.1, the animal dose-response data from the CNT subchronic inhalation studies were limited (and minimally acceptable) for benchmark dose estimation. Only the multistage model provided unique MLE estimates that adequately fit the data. Yet, the rat 10% BMD(L) estimates are similar to the NOAEL and LOAEL estimates, indicating that the BMD(L) estimates are in reasonable agreement (Table A–12).

In standard risk assessment procedure [US EPA 1994, 2012; Kuempel et al. 2006; Schulte et al. 2010], the BMDL, NOAEL, or LOAEL is the estimated point of departure

* Abbreviation for both BMD and BMDL estimates.

(POD) for extrapolation below the range of the data. This low dose extrapolation may be risk-based (e.g., from the 10% BMDL) by linear or nonlinear modeling (depending on the mode of action information); or in noncancer risk assessment, the POD may be adjusted downward to account for uncertainty in the extrapolation from animals to humans and other factors (Section A.6.3.3).

The risk estimates based on the subchronic and shortterm animal studies (Section A.3) generally indicate that risks less than 10% would be associated with working lifetime exposures below 1 $\mu g/m^3$ (8-hr TWA). This is estimated by either linear or model-based low-dose extrapolation of the rat BMDL estimates (Section A.3.3). Although an acceptably low level of risk for early-stage pulmonary effects has not been established, some of the working lifetime excess risk estimates at 1 $\mu g/m^3$ (8-hr TWA) are less than 10% (e.g., approximately 0.5% to 16% for the slight/mild, grade 2) lung effects in the rat subchronic inhalation studies; 95% UCL estimates) (Table A–8). Given the uncertainty about the earlystage lung effects and the shape of the dose-response relationship, the actual risk could be much lower, even zero (if exposures are below an effect threshold). Alternatively, adjusting the POD downward using standard uncertainty factors, also generally results in estimates of <1 $\mu g/\ m^3$ (8-hr TWA concentration) as the working life-time exposure likely to be without appreciable risk of adverse effects (Section A.6.3.3).

5.3.3. Minor Areas of Uncertainty

(1) Effect level estimation

An effect level is a dose associated with a specified effect (or lack of observed effect). A BMD is a statistical estimate of an effect level [Crump 1984, 1995; US EPA 2012]. BMD estimates have several advantages over NOAEL or LOAEL estimates, including use of all the dose-response data, statistical accounting of sample size and variability, and a standard, risk-based definition of effect level (e.g., 10%) in the low region of the dose-response data. The data to estimate BMD(L)s were limited (see Appendix A), but NIOSH considered these data to be minimally acceptable, and adequate statistical fit was obtained for each data set included in the risk assessment (Appendix A).

NOAELs and LOAELs can also be uncertain, as observation of effects can depend on the dose spacing and the number of animals. For example, a low number of animals in a study can result in an apparent NOAEL due to chance, whereas a larger sample size would have more power to detect an effect. A statistical comparison of the NOAEL and BMD estimates showed that the NOAEL was statistically consistent with the 10% BMD estimate (Section A.6.2). In practice, the effect level estimate had little influence on the risk estimates or REL derivation because the BMD(L) estimates were similar to the LOAEL or NOAEL values (Table A–12).

(2) Impact of route of exposure

Different routes of exposure were used in the CNT animal studies, including intratracheal instillation (IT), pharyngeal aspiration (PA), and inhalation. IT and PA are each single administered doses, followed by approximately 1 to 2 months post-exposure time to determine whether the observed effects were reversible, persistent, or progressive. PA permits the delivery of the substance to the lungs of animals that results in a deposition pattern that is more dispersed than that of IT and therefore more similar to inhalation. Inhalation is the most physiologically relevant route of exposure to workers. A study that compared mouse lung responses to SWCNT by PA or inhalation exposure found qualitatively similar responses, although the inhalation exposure was four times more potent at an estimated equivalent lung dose [Shvedova et al. 2008]. Among the various studies and routes of exposure (Tables A–3 through A–5), no clear differences in exposure and risk estimates are seen for route of exposure (versus variability due to other differences among studies).

(3) Rat lung dose estimation

In the absence of CNT-specific lung models, standard rat and human lung dosimetry models were used to estimate either the deposited or the retained lung dose. These two estimates are considered to represent the upper or lower bounds on the possible lung burden estimates. The effect of assuming deposited dose (no clearance) versus retained dose (normal clearance) resulted in a difference of approximately five-fold. However, the actual estimates are expected to lie within this range. The working lifetime exposure estimates (associated with 10% excess risk of early stage lung effects) were lower based on deposited dose estimates than those based on retained dose estimates (Appendix A, Table A–5).

6. RECOMMENDATIONS

In light of current scientific evidence on the hazard potential of CNT and CNF, appropriate steps should be taken to minimize worker exposure through the development of a risk management program and implementation of an exposure control strategy. Elements of that program should include the following:

1. Control worker exposure to CNT and CNF below 1 μg/m^3 8-hr TWA, respirable fraction (elemental carbon) during a 40-hr work week.
2. Conduct comprehensive exposure assessments (including exposures to other potential hazards) as part of an overall hazard surveillance program.
3. Develop guidelines for selecting, installing, and evaluating engineering controls (e.g., local exhaust ventilation, dust collection systems).
4. Educate and train workers on the recognition of potential exposures and in the use of good work practices in the handling of bulk CNT and CNF, as well as CNT- and CNF-containing materials.
5. Develop procedures for the selection and use of personal protective equipment (i.e., clothing, gloves, respirators).
6. Implement a medical surveillance program for workers potentially exposed to CNT or CNF with conduct of specific medical screening tests when warranted (Section 6.7).
7. Conduct routine (e.g., annual) and systematic evaluation of worker exposure to CNT or CNF when there is a process change in how CNT or CNF are manufactured or handled.
8. Encourage workers to wash hands before eating, smoking, or leaving the worksite.

9. Establish facilities for showering and changing clothes, with separate facilities for storage of non-work clothing, to prevent the inadvertent cross contamination of other areas (including take-home).

6.1. Exposure Assessment

NIOSH is recommending that a respirable massbased airborne concentration measurement be used to monitor worker exposure to all types of CNT and CNF until additional data are available to determine whether other measurement metrics or techniques would be more effective in protecting workers' health. NIOSH is currently evaluating the efficacy of various sampling techniques for measuring CNT and CNF and may make additional recommendations at a later date.

Personal exposure concentrations to CNT and CNF can be determined as elemental carbon (EC) by NIOSH Method 5040 [NIOSH 1994; Birch 2004a, b]. Whenever possible, a bulk sample of the CNT/ CNF material should be analyzed to establish the thermal profile for the material(s) (Appendix C). Measurement results from NIOSH Method 5040 should provide a reasonable estimate of a worker's respirable exposure to CNT and CNF at the NIOSH REL of 1 $\mu g/m^3$ 8-hr TWA when the predominant workplace exposure to EC material is CNT or CNF.

6.1.1. Exposure Monitoring Program

An exposure-monitoring program should be established to ensure that worker exposures to CNT and CNF are maintained below the REL. The program should consist of a plan designed to do the following: (1) characterize exposures of all exposed workers; (2) identify sources of potential EC exposures (e.g., diesel soot, carbon black) that may interfere with the interpretation of worker CNT and CNF exposures; (3) identify specific work areas or job tasks where worker exposures exceed or may exceed the REL; and (4) assess the effectiveness of engineering controls, work practices, PPE, training, and other factors used in reducing airborne exposures. To implement the plan an exposure assessment strategy should be developed. The details of the strategy will depend on a number of factors, including the number of workers potentially exposed to CNT or CNF and the day-to-day and worker-to-worker variability in airborne concentrations.

An important first step in applying any strategy is to develop an inventory of the processes and job activities (e.g., handling of dry powders, use of composite materials) that place workers at risk of exposure. This inventory can be used to determine the number of workers potentially exposed and a qualitative assessment as to the workers and processes with the highest potential for exposure.

The strategy should also incorporate provisions to quantify the airborne release of CNT and CNF occurring at specific processes or job activities to provide "activity pattern data" [Duan and Mage 1997]. Activity pattern data are useful for identifying possible causes of high exposure for remediation; however, these data are vulnerable to spatial variation in exposure concentrations and should not be used in predicting worker exposures. For example, in the study by Birch et al. [2011b], personal exposure to CNF was much higher than area samples, depending on location. Respirable EC exposure for two employees working mainly in a thermal treatment area was approximately 45 $\mu g/m^3$, and a CNF reactor area was about

80 μg/m^3, while the corresponding area samples were about 32 μg/m^3 in the thermal treatment area and 13 μg/m^3 in the CNF reactor area. The EC concentration in the reactor area was less than half that in the thermal treatment area, but the personal sample collected in the reactor area was nearly twice as high. Because area samples are often not predictive of personal exposure, extrapolating personal exposure from area concentrations should not be done without a thorough assessment of the workplace to establish whether a valid extrapolation is possible [Birch et al. 2011b]. NIOSH [NIOSH 2009a] and others [Brouwer et al. 2009; Methner et al. 2010a; Ramachandran et al. 2011] have developed exposure assessment guidance for determining the release of engineered nanoparticles that can be adapted for determining sources of exposure to CNT and CNF.

To ensure that worker exposure to CNT or CNF is being maintained below the REL, several exposure measurement strategies are available [NIOSH 1977; Corn and Esmen 1979; Leidel and Busch 1994; Rappaport et al. 1995; Lyles et al. 1997; Bullock and Ignacio 2006]. These strategies can be tailored to the specific workplace depending on the number of workers, complexity of the work environment (e.g., process type and rate of operation, exposure control methods, physical state and properties of material) and available resources. One approach for determining worker exposure would be to initially target similarly exposed groups of workers [Corn and Esmen 1979; Leidel and Busch 1994]. This initial sampling effort may be more time efficient and require fewer resources for identifying workers with exposures to CNT or CNF above the REL. However, this measurement strategy may produce incomplete and upwardly biased exposure estimates if the exposures are highly variable [Kromhout 2009]. Therefore, repeated measurements on randomly selected workers may be required to account for between- and within-worker variation in exposure concentrations [Rappaport et al. 1995; Lyles et al. 1997]. Because there is no 'best' exposure measurement strategy that can be applied to all workplaces, multiday random sampling of workers (all workers, if the exposed workforce is small) may be required to have an accurate assessment of worker airborne exposure concentrations to CNT and CNF.

6.1.2. CNT and CNF Measurement

A multi-tiered exposure measurement strategy is recommended for determining worker exposure to CNT and CNF (see Figure 6–1). The selection of workers and the frequency in which they should be sampled should follow guidelines established for the exposure monitoring program (Section 6.1.1).

As part of the evaluation of worker exposures to CNT and CNF, 'background' samples for EC determination should be collected outdoors (or at air intakes of the facility) and at indoor locations where exposure to CNT or CNF is unlikely. The EC concentrations (using Method 5040) determined from 'background' samples should be subtracted from the EC personal sample results to determine whether worker exposures exceeded the REL. Initially, more samples may be required to characterize the workplace thoroughly. This initial assessment will help refine the sampling approach and determine whether EC interference is an issue. Careful consideration of environmental background is essential. For example, outdoor EC may sometimes be higher than indoor background depending on the facility's air handling system. If so, the indoor EC background may be more representative of area and worker samples.

In workplaces where exposure to other types of EC (e.g., diesel soot, carbon black) may occur, the initial evaluation of a worker's exposure should include the simultaneous collection of a personal respirable EC sample and a personal sample for electron microscopy analysis (e.g., TEM, SEM). Electron microscopy analysis, in conjunction with energy dispersive x-ray spectroscopy (EDS), can be used for CNT and CNF identification. In addition, consideration should be given to the sizing and counting of CNT and CNF structures during electron microscopy analysis should future efforts to control occupational exposures be based on a different exposure metric (e.g., number concentrations of airborne CNT and CNF structures in a given size bin). While no specific electron microscopy (e.g., TEM, SEM) method exists for the sizing and counting of CNT and CNF structures, methods used in the analysis of other 'fibrous' materials are available [NIOSH 1994a; ISO 1999, 2002] and could be adapted in the characterization of exposures.

NIOSH investigators have conducted a number of surveys at CNT and CNF producers and/or secondary users [Evans et al. 2010; Birch 2011a; Birch et al. 2011b; Dahm et al. 2011]. In many cases 'background' EC concentrations were <1 μg/m^3. In the study reported by Dahm et al. [2011], results for PBZ samples were often "non-detect" or between the LOD and LOQ (i.e., semi quantitative results) for Method 5040. In these facilities [Dahm et al. 2011], only small amounts of material were being handled and the tasks were over a short duration. Thus, time-weighted air concentrations determined over extended periods were low. These data indicate that in workplaces where airborne exposures to CNT and CNF are expected to be low, or where worker exposure may be sporadic or shortterm, a higher flow rate (e.g., 4 lpm) respirable dust sampler (cyclone) and a 25-mm filter should be used to increase the amount of sample collected for quantification.

At job tasks where dusts are generated during cutting, sanding, or grinding of CNT/CNF polymer composites, the accuracy of determining the EC fraction of the dust can vary depending on the polymer and sample loading (Appendix C). Depending on the dust concentration, the sample can be easily overloaded with organic carbon (OC) because of the high relative OC content of the polymer. Due to OC overload and/or polymer pyrolysis, it may not be possible to quantify exposures to CNT and CNF by Method 5040 if they are incorporated as powders in composite operations. Further, even if the EC fraction of a composite dust can be determined with reasonable accuracy, both unbound and polymer-bound CNT and CNF are determined by Method 5040, if present. Also, in a polymer composite particle, the CNT/CNF is usually bound within the polymer (or resin) matrix, dissimilar to a particle of unbound material. Analysis of samples by electron microscopy may be helpful in identifying the presence of matrix bound and unbound CNT or CNF and for determining a course of action in controlling exposures. An effort to improve the analysis of samples containing dusts of polymer composites is ongoing.

Metals employed as catalysts in the synthesis of CNT/CNF also were considered by NIOSH as potential markers of CNT/CNF airborne exposure, but purified CNT and CNF have low metal contents (1% or less by weight). Detection limits for inductively coupled plasma atomic emission spectroscopy (ICP-AES) may not be adequate at low CNT/CNF concentrations [Birch et al. 2011b]. Inductively coupled plasma mass spectrometry (ICP-MS) offers detection limits superior to ICP-AES and may be useful, depending on the amount of CNT/CNF collected, metal and percent metal con- tent, and whether other aerosol sources are present that would interfere with analysis. However, if a metal is employed as a surrogate measure of CNT/ CNF, minimal background interference and correlation with CNT/CNF mass (or other relevant metric) would be required [Birch et al. 2011b]. Iron was not a useful indicator of CNF exposure in a study reported by Birch et al. [2011b]. There was no correlation between the iron and CNF concentrations found in a CNF manufacturing facility, because the major iron source was not CNF-derived. In addition, the LOD for ICP-AES was not adequate for its determination at low EC concentrations (e.g., near the EC LOQ).

6.1.3. Method 5040 Limit of Detection

As with all analytical methods, the LOD is a varying number. However, the airborne EC LOD originally reported for NIOSH Method 5040 (i.e., about 2 $\mu g/m^3$), or an LOQ of 7 $\mu g/m^3$ was a high estimate [NIOSH 2010]. The LOD was based on analysis of precleaned media blanks from different filter lots, during a 6-month period, and by different analysts at two different laboratories. Further, variability for the total carbon (TC) results was used to estimate the LOD rather than EC results. These combined factors gave a conservative (high) estimate of the EC LOD.

In practice, a much lower EC LOD is obtained by NIOSH Method 5040 than was originally reported, because the variability for EC results for a set of media blanks submitted (with the sample set) for the LOD (LOQ) determination is much lower than reported for the total carbon (TC) results. Thus, if EC is of primary interest, as with CNT/CNF measurement, and the level of organic carbon (OC) contamination is acceptable (with respect to the OC and TC LOD), EC results for as-received filters should be used to determine the EC LOD (Appendix C).

Estimates of the EC LODs and LOQs (in units μg EC/cm^2 of air) determined with 25-mm and 37-mm quartz filter media from a given lot, and with manual splits assigned are reported in Table 6–1. OC-EC splits for the media blanks were assigned at the point when oxygen is introduced so the baseline signal is integrated over the region in which EC is removed (oxidized) from the filter.

Because there are many possible OC sources, and bulk CNT/CNF contain little OC, EC is a better indicator of exposure than TC. Nevertheless, high particulate OC concentrations indicate air contamination, and these data can be useful for general industrial hygiene purposes if care is taken to correct for OC media contamination (Appendix C). Unlike EC, OC contamination (e.g., through contact with a contaminated surface and/or vapor adsorption) of the quartz filter media is common. Consequently, the OC (and TC) LOD is higher than for EC, and the OC (and TC) results may have significant positive bias. Bias is especially apparent when the particulate OC air concentrations and sampled air volumes are low. To obtain a more accurate estimate of the particulate OC air concentration, an OC blank correction should be applied. Blank correction can be accomplished by subtracting the OC media blank or (preferably) by a tandem filter correction (organic carbon sampling artifacts section in Appendix C), with the latter generally being more accurate. Mean OC blanks, LODs, and LOQs for 25-mm and 37-mm quartz filter media are reported in Table 6–2 (units are μg OC/cm^2).

Two additional sets (n = 10 for each set of five filters) of 37-mm filters were analyzed several months apart in 2010. The pooled EC results are comparable to those obtained previously. Results (μg C/cm^2), including OC results for both sets and TC results for one are given in Table 6–3. The results in Table 6–3 represent duplicates on two sets of 5 filters.

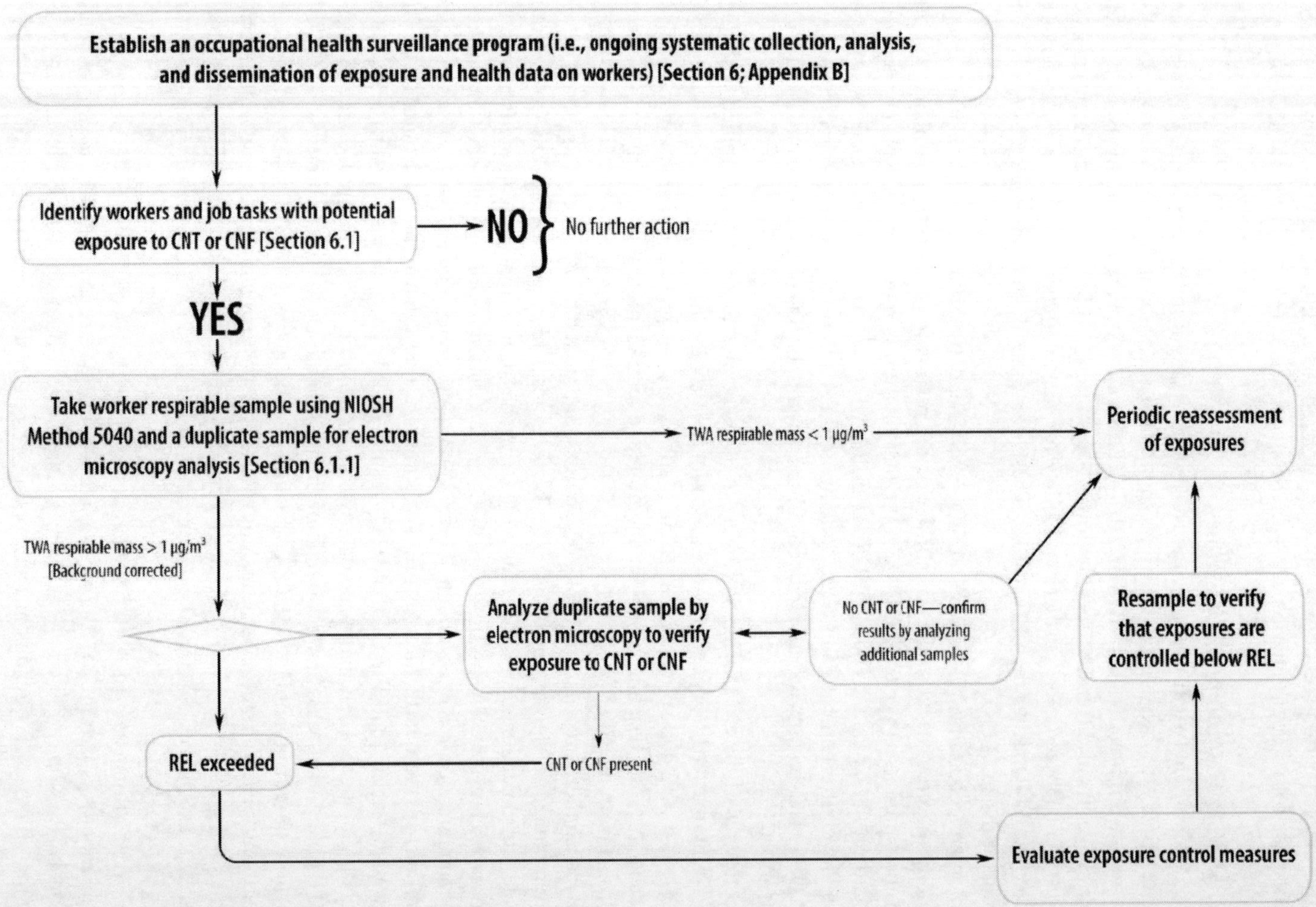

Figure 6–1. Exposure measurement strategy for CNT and CNF.

Table 6–1. EC LODs and LOQs for 25- and 37-mm filters (ug EC/cm²)

25-mm filter	EC (n = 10)	37-mm filter	EC (n = 6)
Mean	0.063	Mean	0.033
Standard deviation	0.030	Standard deviation	0.028
LOD	0.09	LOD	0.08
LOQ	0.30	LOQ	0.28

Table 6–2. OC LODs and LOQs for 25- and 37-mm filters (ug OC/cm²)

25-mm filter	OC (n = 10)	37-mm filter	OC (n = 6)
Mean	1.41	Mean	1.94
Standard deviation	0.413	Standard deviation	0.281
LOD	1.24	LOD	0.84
LOQ	4.13	LOQ	2.81

Table 6–3. OC, EC, and TC LODs and LOQs for 37-mm filters (ug C/cm²)

37-mm filter	OC	EC	OC	EC	TC
Mean	1.31	0.03	1.44	0.03	1.52
Standard deviation	0.164	0.016	0.304	0.024	0.318
LOD	0.49	0.05	0.91	0.07	0.95
LOQ	1.64	0.16	3.04	0.24	3.18

As stated on previous page, NIOSH Method 5040 LOD depends on the air volume, filter size, sample portion analyzed (usually 1.5 cm^2), and the media blank variability. The latter is used to determine the LOD in unit's µg/cm^2. Expressed as an air concentration, the EC LOD (µg/m^3) corresponding to the EC LOD (µg/cm^2) determined with media blanks (i.e., LOD = 3 times the standard deviation for the blanks) can be calculated by the following equation:

$$EC\ LOD\left(\mu g/m^3\right) = \frac{EC\ LOD\left(\mu g/cm^2\right) \times deposit\ area\left(cm^2\right)}{air\ volume\left(m^3\right)}$$

This equation explains why a lower LOD (µg/m^3) can be obtained by reducing the filter size (deposit area), increasing the air volume, and minimizing the variability for the media blanks (i.e., the EC LOD in µg/cm^2). The LOD is improved using a smaller filter size since the deposit density is higher for an equivalent mass deposited. The same applies to the LOQ, commonly defined as 10 times the standard deviation (SD) for the blanks, or 3.3 times the LOD.

If 0.02 µg EC/cm^2 is taken as the SD for media blanks (with manual OC-EC split adjustment), the LODs and LOQs (in µg EC/m^3) for different air volumes, 25-mm and 37-mm filters, and a 1.5 cm^2 filter portion analyzed would be as listed in Table 6–4. Results for a SD double this value (i.e., SD = 0.04 µg EC/cm^2) also are reported as worst-case estimates, but are seldom this high. If SDs for media blanks are frequently above 0.02 ug EC/cm^2) the cause of the high blank variability should be identified and corrected.

Based on EC results for media blanks (Tables 6–1 and 6–3), a filter loading of about 0.3 $\mu g/cm^2$ (i.e., at or above the EC LOQs reported in Tables 6–1 and 6–3) will provide quantitative results.

As described above (6.1.2), higher flow rate respirable samplers (cyclones) with a 25 mm cassette can improve the sample collection to permit measurement of CNT and CNF above the LOQ (i.e., 1 $\mu g/m^3$) for samples collected for less than full work shift. Examples of sampling periods and flow rates that provide air volumes of about a halfcubic meter or higher (shaded area in table below) are listed in Table 6–5. A larger filter portion also can be used to further lower the LOD, but the instrument's small (quartz tube) oven and need for proper sample alignment limit the amount of sample that can be analyzed.

6.2. Engineering Controls

One of the best ways to prevent adverse health effects from exposure to CNT and CNF is to eliminate exposure and minimize risks early in the design or re-design of manufacturing and down-stream user processes (see NIOSH prevention through design (PtD) at: www.cdc.gov/niosh/topics/PtD/). This can be accomplished through the establishment of a process safety management (PSM) program. PSM entails the development and implementation of programs or systems to ensure that the practices and equipment used in potentially hazardous processes are adequate and appropriately maintained. An integral part of the PSM program is the conduct of a process hazard analysis prior to the initiation of work to identify where sources of exposure to CNT or CNF may occur so that process equipment can be designed or re-designed to minimize the risk of exposure. At a minimum, the elements of the PSM program should be consistent with those required in the OSHA Process Safety Management Standard [29 CFR 1910.119].

In workplaces where CNT or CNF can't be substituted with a less hazardous or nonhazardous material then all process equipment and other equipment involved with the handling of CNT and CNF should incorporate the necessary engineering control measures to prevent worker exposure to CNT and CNF. Because of limited published workplace exposure data for CNT and CNF, it is unknown whether worker respirable mass exposures to CNT and CNF can be maintained at all workplaces below the NIOSH REL of 1 $\mu g/m^3$ EC as an 8-hour TWA. However, exposure control techniques such as source enclosure (i.e., isolating the generation source from the worker) and well-designed local exhaust ventilation (LEV) systems equipped with high efficiency particulate air (HEPA) filters have been shown to be effective for capturing airborne nanoparticles including CNT and CNF [Old and Methner 2008; NIOSH 2009a; Evans et al. 2010]. A general description of exposure control techniques and their advantages and disadvantages is given in Table 6–6. The selection of the exposure control technique should take into account the quantity and physical form of the nanomaterial (e.g., dispersible powder, liquid slurry, contained in a matrix) and the task duration and frequency in which workers come into contact with the material (Table 6–7). For instance, working with materials containing CNT or CNF (e.g., encapsulated in a solid) may require a different type of an exposure control system than would be required for large quantities of CNT and CNF in a highly dispersed free form. Processes involved in the cutting, grinding, or drilling of solid materials containing CNT or CNF should incorporate appropriate engineering controls (e.g., local exhaust ventilation) to prevent aerosol release, whereas the manufacturing (i.e., product collection at reactor) and handling of dry bulk CNT or CNF should be performed in enclosed, and when warranted, HEPA-ventilated systems. HEPA filtration has been shown to be effective in capturing nanoscale particles and should be considered in situations where emissions may be regular, where processes are repeated, and where higher quantities are used in a way that may lead to emissions. The handling of research quantities of CNT and CNF in laboratories is best performed using a laboratory fume hood, such as a low-flow or air-curtain hood [Tsai et al. 2010], or use of a glove box to minimize worker exposure [NIOSH 2012]. All exposure control systems should be properly designed, tested, and routinely maintained to ensure maximum efficiency [ACGIH 2007].

6.3. Worker Education and Training

Establishing a program that includes the education and training of workers on the potential hazards of CNT and CNF and their safe handling is critical to preventing adverse health effects from exposure. Research has shown that training can attain immediate and longterm objectives when (1) workers are educated about the potential hazards of their job, (2) there are improvements in knowledge and work practices, (3) workers are provided the necessary skills to perform their job safely, and (4) there is management commitment and support for work-place safety [NIOSH 2010b]. The requirements for the education and training of workers as specified in the *OSHA Hazard Communication Standard* (29 CFR 1910.1200), the *Hazardous Waste Operation and Emergency Response Standard* (29 CFR 1910.120), and as described by Kulinowski and Lippy [2011] for workers exposed to nanomaterials, provide a minimum set of guidelines that can be used for establishing an education and training program. The establishment of a program should have written procedures (e.g., standard operating procedures [SOPs]) for: (a) ensuring management commitment to control exposures, (b) identifying and communicating potential hazards to workers, (c) assessing workplace exposures to CNT and CNF, (d) identifying and implementing engineering and work practice controls, (e) establishing documentation of risk management actions taken, and (f) periodically reviewing the adequacy of controls and other preventive practices. Management should systematically review and update these procedures and convey to workers actions taken to resolve and/or improve workplace conditions.

Table 6–4. EC LODs and LOQs estimated with media blanks

SD blank (μg EC/cm^2)	Limit	EC limit (μg/cm^2)	EC LOD and LOQ (μg EC/m^2)					
			3 m^3 air		1 m^3 air		0.5 m^3 air	
			37-mm filter	25-mm filter	37-mm filter	25-mm filter	37-mm filter	25-mm filter
0.02	LOD	0.06	0.17	0.07	0.51	0.21	1.02	0.42
	LOQ	0.20	0.57	0.23	1.70	0.69	3.40	1.38
0.04	LOD	0.12	0.34	0.14	1.02	0.42	2.04	0.83
	LOQ	0.40	1.13	0.46	3.40	1.38	6.80	2.77

Table 6–5. Examples of sampling periods and flow rates

Flow rate (lpm)[†]	Air volumes (m^3) for indicated sampling periods (hours) and flow rates			
	1 h[‡]	2 h	4 h	8 h
2	120	240	480	960
4	240	480	960	1920
6	360	720	1440	2880
7	420	840	1680	3360

† Liters per minute (lpm).

‡ Sampling period, hours, and highest flow rate tested at NIOSH laboratory. Tested with a Leland Legacy pump and 25-mm quartz-fiber filter.

Table 6–6. Examples of engineering controls

Containment category and description	Advantages	Disadvantages
A. Dilution ventilation and no engineering controls Supply and exhaust large volumes of air (typically > 10 air changes/hr.) throughout the work area to dilute airborne emissions. **For general facility HVAC needs. *Not recommended* for controlling worker exposure to CNT and CNF.**	No local exhaust ventilation (LEV) or equipment enclosures required. Disperses/dilutes airborne emissions throughout work area.	Does not control exposure at the source, spreads emissions throughout work area potentially exposing other workers. Often requires large airflow exhausts to dilute contaminants to below OEL increasing operating costs. Should only be considered when contaminant generation is reasonably uniform and toxicity of material is low.
B. Local exhaust Ventilation (LEV) Hoods or enclosures on process equipment that exhaust air at the emission source to collection equipment and away from the worker's breathing zone. Includes: B.1. Laboratory fume hoods (typically 80–120 ft/min face velocity) with HEPA filter B.2. Biological safety cabinet Class II B.3. LEV incorporated at source of exposure and can be built into hand-held tools	Capture emissions at their source with well-designed hoods. Hoods can be tailored to the process or work task to optimize the capture of emissions. Usually requires less overall exhaust airflow rates than dilution ventilation systems.	Air volumes and face velocity of LEV must be maintained to ensure the capture of emissions. Workers must be trained in the correct use. Fume hood sash opening needs to be adjusted to ensure proper hood face velocity. System exhaust flow rate may need careful evaluation to ensure adequate capture while minimizing loss of product.
C. Down flow booths Small room or enclosure with low velocity (100 ft/min) downward airflow to push/pull contaminants away from the worker's breathing zone.	Emissions pushed away from the worker's breathing zone. Flexible control that can be used for several tasks/ operations. Useful for manual operations for which a more contained enclosure is not feasible (e.g., larger amounts of materials or equipment).	Air volumes and control velocities of booth must be monitored/maintained to ensure proper performance. Worker technique and interface with the work process can interfere with the capture of emissions. Workers must be trained in the correct use.

Containment category and description	Advantages	Disadvantages
D. Closed process design (isolation) All steps of the process or job task are sealed with little chance of worker exposure. Examples: D.1. Glove box isolators (with HEPA filtered exhaust) D.2. Biological safety cabinet class III	Emission source confined. Minimizes external contamination. Need for worker PPE (e.g., respirator) reduced.	More time required moving materials and equipment in and out of enclosure. Difficulty in manipulating materials when wearing gloves. Limitations on size of material that can be placed inside of an isolator. Need to periodically clean the enclosure.

Adapted from Industrial Ventilation [ACGIH 2007].

Table 6–7. Engineering controls to reduce CNT and CNF exposures

Process/activity	Potential exposure source and recommended containment of exposure*
A. Pilot and research development operations	**Exposure Source:** Synthesis of CNT and CNF by fluidized-bed, chemical vapor deposition, etc.: a) collection/harvesting after synthesis, b) powder transfer, c) cleaning reactor, d) removal of CNT and CNF from a substrate, e) purification and/or functionalization of CNT or CNF [note: potential exposures are generally to *small quantities* of CNT and CNF (i.e., μg, mg) compared to exposure to larger amounts (e.g., kg) during full-scale manufacturing/synthesis (see C below)]. **Exposure Controls:** a) laboratory fume hood (with HEPA filtered exhaust when warranted), b) HEPA-filtered exhausted enclosure (glove box), or c) biological safety cabinet. Local exhaust ventilation (LEV) may be required when opening reactor and during harvesting.
B. Research laboratories	**Exposure Source:** Handling (e.g., mixing, weighing, blending, transferring) *small quantities (e.g., μg, mg) of CNT or CNF powder or during sonication* of a CNT or CNF liquid suspension. **Exposure Controls:** a) laboratory fume hood (with HEPA filtered exhaust when warranted), b) HEPA-filtered exhausted enclosure (glove box isolator), or c) biological safety cabinet.
C. CNT and CNF manufacturing and synthesis	**Exposure Source:** Synthesis of CNT and CNF by fluidized-bed, chemical vapor deposition, etc., including: a) collection/harvesting after synthesis, b) drum and bag filling, c) powder transfer, d) cleaning reactor, e) removal of CNT and CNF from a substrate, f) purification and/or functionalization of CNT or CNF [Note: potential exposures are generally to *large* quantities (e.g., kg) of CNT and CNF]. **Exposure Controls:** Dedicated ventilated room with HEPA filtered exhaust, and/or LEV at source of exposure with HEPA filtered exhaust. Examples: ventilated bagging/weighing station and/or laminar down-flow booth or non-ventilation options such as continuous liner off-loading systems for bagging operations. Ventilated bag dumping stations for product transfer.
Production and use of CNT and CNF enabled materials and composites	**Exposure Source:** Mixing, weighing, and transferring of *small quantities* of CNT or CNF powder or liquid suspension, including the: a) incorporation of CNT or CNF into matrices (e.g., polymer composites) and into coatings (e.g., inks) and, b) spraying CNT or CNF on surfaces.

Table 6–7. (Continued)

Process/activity	Potential exposure source and recommended containment of exposure*
	Exposure Controls: a) Laboratory fume hood (with HEPA filtered exhaust when warranted), b) HEPA-filtered exhausted enclosure (glove box isolator), or c) biological safety cabinet. **Exposure Source:** Handling *large quantities* of CNT or CNF powder that involves pouring and blending into other matrices. In addition, spinning, twisting, weaving of CNT into making rope, cloth, etc.; spray coating of surfaces. **Exposure Controls:** Isolation techniques such as a dedicated ventilated room or process enclosure with HEPA filtered exhaust. Process-based controls such as ventilated bagging/weighing station, laminar down-flow booth or non-ventilation options such as continuous liner off-loading systems for bagging operations. Ventilated bag dumping stations for product transfer. **Exposure Source:** Grinding, sanding, cutting, drilling or other mechanical energy applied to enabled-materials/composites containing CNT or CNF. **Exposure Controls:** For the handling of *small pieces* of CNT or CNF enabled materials/composites: a) laboratory fume hood (with HEPA filtered exhaust when warranted), b) HEPA filtered exhausted enclosure (glove box isolator), or c) biological safety cabinet. **Exposure Controls:** For handling *large* CNT or CNF enabled materials/ composites and where use of isolation techniques such as large ventilated enclosures are not feasible: a) use LEV at exposure source with HEPA filtered exhaust (may include LEV built into a hand-held tool), b) ventilated down-flow booths with HEPA filtered exhaust, c) laboratory fume hood (with HEPA filtered exhaust) and/or d) wet dust suppression machining techniques such as wet saws (if applicable).

*Note: Factors that influence selection of appropriate engineering controls and other exposure control strategies include the physical form (e.g., dry dispersible powder, liquid slurry, in a matrix/composite), task duration, frequency, and quantity of CNT or CNF handled. Measurement of airborne exposure at the potential source of emission should be performed to confirm the effectiveness of the control measure.

A program for educating workers should also include both instruction and "hands-on" training that addresses the following:

- The potential health risks associated with exposure to CNT and CNF.
- The safe handling of CNT, CNF, and CNT- and CNF-containing materials to minimize the likelihood of inhalation exposure and skin contact, including the proper use of engineering controls, PPE (e.g., respirators, gloves), and good work practices.

6.4. Cleanup and Disposal

Procedures should be developed to protect workers from exposure to CNT and CNF during the cleanup of CNT and CNF spills and CNT- or CNF-contaminated surfaces. Inhalation and dermal exposures will likely present the greatest risks. The potential for inhalation exposure during cleanup will be influenced by the likelihood of CNT and CNF becoming airborne, with bulk CNT and CNF (powder form) presenting a greater inhalation potential than CNT and CNF in solution (liquid form), and liquids in turn presenting a greater potential risk than CNT- and CNF-encapsulated materials.

It would be prudent to base strategies for dealing with spills and contaminated surfaces on the use of current good practices, together with available information on exposure risks. Standard approaches for cleaning powder spills can be used for cleaning surfaces contaminated with CNT or CNF. These include using HEPA-filtered vacuum cleaners, wiping up CNT and CNF (powder form) using damp cloths, or wetting the powder before wiping. Liquid spills containing CNT or CNF can typically be cleaned by applying absorbent materials/liquid traps. If vacuum cleaning is employed, care should be taken that HEPA filters are installed properly and bags and filters changed according to manufacturer's recommendations. Dry sweeping or air hoses should not be used to clean work areas.

The handling and disposal of waste (including all cleaning materials) and other contaminated materials (e.g., gloves) should comply with all applicable regulations (e.g., federal, state, local).

6.5. Personal Protective Clothing

There are no regulations or guidelines for the selection of protective clothing or other apparel against exposure to CNT and CNF; however, the Occupational Safety and Health Administration (OSHA) requires employers to provide employees with hand protection when exposed to hazards [OSHA 1910.138(a)]. Currently, limited information is available to assess the exposure and health hazards of skin exposure to CNT and CNF. In a study to determine potential airborne and dermal exposures to SWCNT during manufacturing and handling, workers' dermal exposure was estimated by placing cotton gloves over the rubber gloves used by workers [Maynard et al. 2004]. Dermal exposure estimates for SWCNT on individual gloves (total hand area) ranged from 217 μg to 6020 μg, with most of the SWCNT material appearing on the parts of the gloves in direct contact with surfaces. Results from experimental studies with various types of nanoparticles found that dermal penetration of nanoparticles may occur under certain conditions of exposure (e.g., flexing of skin) [Ryman-Rasmussen et al. 2006; Rouse et al. 2007] and that factors such as size, shape, water solubility, and surface coating directly affect a nanoparticle's potential to penetrate the skin [Sayes et al. 2004; Ryman-Rasmussen et al. 2006]. The results from *in vitro* studies, using primary or cultured human skin cells and engineered human skin, show that SWCNT and MWCNT are able to enter cells and cause the release of pro-inflammatory cytokines, induce free radical generation and oxidative stress, and decrease cell viability [Shvedova et al. 2003; Monteiro-Riviere et al. 2005; Murray et al. 2009; Vankoningsloo et al. 2010]. Vankoningsloo et al. [2010] also found the surface properties of MWCNT played a determinant role in their interaction with cells, and when nanotube agglomeration was decreased, there was an increase in cytotoxicity on keratinocytes. The topical application of SWCNT (160 μg) to SKH-1 mice caused inflammation that was localized around or within the hair follicles; however, no significant changes were observed at the lowest dose (40 μg) tested [Murray et al. 2009]. It was concluded that topical exposure to unpurified SWCNT at doses > 80 μg/mouse are capable of inducing free radical generation, oxidative stress, and inflammation. However, the results of dermal toxicity testing with one type of MWCNT (Baytubes®) found no evidence of acute skin irritation or sensitization and only mild eye irritation in rabbits when tested according to OECD test guidelines [Pauluhn 2010b].

Given the limited amount of data on dermal exposure to CNT and CNF, it would be prudent to wear protective clothing and gloves when

- all technical measures to eliminate or control the release of exposure to CNT and CNF have not been successful, or
- in emergencies.

If protective clothing and/or gloves are worn, particular attention should be given to preventing CNT and CNF exposure to abraded or lacerated skin. Based on limited experimental evidence, airtight fabrics made of nonwoven textile seem to be more efficient in protecting workers against nanoparticles than fabrics made of woven cotton or polyester [Golanski et al. 2009; Golanski et al. 2010]. The results of a study designed to evaluate the penetration of nano- and submicron particle penetration through various nonwoven fabrics found minimal penetration (< 5%) of iron oxide particles (< 100 nm) through nonwoven fabrics typically used for hospital frocks, hoodless coveralls, and firefighter ensemble insulation [Gao et al. 2011]. The challenge when selecting appropriate protective apparel is to strike a balance between comfort and protection. Garments that provide the highest level of protection (e.g., an impermeable Level A suit) are also the least comfortable to wear for long periods of time, while garments that are probably the least protective (e.g., thin cotton lab coat) are the most breathable and comfortable to wear. The efficiency of commercial gloves to prevent dermal exposure to nanoparticles varies depending on the glove material, its thickness, and the manner in which it is used (e.g., long exposure times, other chemical exposures) [NanoSafe 2008; Golanski et al. 2009, 2010]. The proper selection of gloves should take into account the resistance of the glove to the chemical attack by both the nanomaterial and, if suspended in liquids, the liquid [USDOE 2007]. If protective gloves (e.g., nitrile, neoprene, latex) are used then "double gloving" may be needed when the worker requires physical protection (e.g., working with sharp instruments) in addition to chemical protection. Special attention should also be given to the proper removal and disposal of contaminated gloves to prevent skin contamination. Gloves should also be visually inspected for tears and routinely replaced.

6.6. Respirators

When engineering controls and work practices cannot reduce worker CNT and CNF exposures to below the REL, then workers should be provided respiratory protection. The use of respirators may also be advisable for certain work tasks that place workers at risk of potentially high peak concentrations of CNT and CNF (e.g., the cleanup of CNT and CNF spills or debris, maintenance of equipment used to process CNT- and CNF-materials, the cleaning or disposal of filtration systems used to capture CNT and CNF aerosols). The OSHA respiratory protection standard (29 CFR 1910.134) sets out the elements of a respirator program for both voluntary and required respirator use. When respirators are provided for worker protection, the OSHA respiratory protection standard requires that a respiratory protection program be established 29 CFR 1910 (c)(1). Elements of the program include (1) a medical evaluation of the worker's ability to perform the work while wearing a respirator; (2) regular training of personnel; (3) periodic workplace exposure monitoring; (4) procedures for selecting respirators; (5) respirator fittesting; and (6) respirator maintenance, inspection, cleaning, and storage. The effectiveness of the program should be evaluated regularly and respirators should be selected by the person who is in charge of the program and knowledgeable about the work-place and the limitations associated with each type of respirator. The voluntary use of respirators are permitted, but must comply with the provisions set forth in CFR 1910.134(c)(2)(i) and 1910.134(c)(2)(ii).

Based on published workplace monitoring data for CNT [Maynard et al. 2004; Han et al. 2008a; Bello et al. 2008, 2009, 2010; Dahm et al. 2011] and CNF [Methner et al. 2007; Evans et al. 2010], a NIOSH-approved filtering facepiece respirator, or elastomeric half-facepiece particulate respirator equipped with a 95 or 100 series filter, should provide adequate protection when properly fit-tested on the worker [Shaffer and Rengasamy 2009], and where engineered controls have been installed to reduce exposures. A properly fittested, half-facepiece particulate respirator or a filtering facepiece respirator will provide protection at exposure concentrations up to 10 times the REL. Other classes of respirators are available that provide a higher level of protection (Table 6–8). The publication *NIOSH Respirator Selection Logic 2004* provides guidance for selecting an appropriate respirator [NIOSH 2005].

When selecting the appropriate respirator, the respirator program manager should consider the particle size in which workers will be potentially exposed [Rengasamy and Eimer 2011], and the presence of other workplace aerosols. Based on this information, the respirator program manager may decide to choose a respirator with a higher assigned protection factor (APF) or choose a respirator with a higher level of filtration performance (e.g., changing from an N95 to a P100). Studies on the filtration performance of N-95 filtering facepiece respirators have found that the mean penetration levels for 40 nm particles range from 1.4% to 5.2%, indicating that 95 and higher performing respirator filters would be effective at capturing airborne CNT and CNF [Bałazy et al. 2006; Rengasamy et al. 2007, 2008]. Recent studies also show that nanoparticles <20 nm are also effectively captured by NIOSH-approved filtering facepiece respirators as predicted by the single fiber theory [Rengasamy et al. 2008, 2009].

6.7. Medical Screening and Surveillance

The toxicological evidence summarized in this document leads to the conclusion that workers occupationally exposed to CNT and CNF may be at risk of adverse respiratory effects. These workers may benefit from inclusion in a medical screening and surveillance program recommended to help protect their health (Figure 6–2) [NIOSH 2009b].

6.7.1. Worker Participation

Workers who could receive the greatest benefit from medical screening include the following:

- Workers exposed to concentrations of CNT or CNF in excess of the REL (i.e., workers exposed to airborne CNT or CNF at concentrations above 1 µg/m^3 EC as an 8-hr TWA).
- Workers in areas or in jobs who are qualitatively determined (by the person charged with program oversight) to have the potential for exposure to intermittent elevated airborne concentrations of CNT or CNF (i.e., workers are at risk of being exposed if they are involved in the transfer, weighing, blending, or mixing of bulk CNT or CNF; or in the cutting, grinding, or drilling of composite materials containing CNT or CNF; or in areas where such activities are carried out by others).

Table 6–8. Respiratory protection for exposure to CNT and CNF

Airborne concentrations of CNT and CNF or conditions of use* options	Minimum respiratory protection
1–10 µg/m^3 (10 × REL)	Any filtering facepiece respirator or air-purifying, elastomeric half-facepiece respirator equipped with appropriate type of particulate filter† Any negative pressure (demand), supplied-air respirator equipped with a half-mask
≤ 25 µg/m^3 (25 × REL)	Any powered, air-purifying respirator equipped with a hood or helmet and a high-efficiency particulate air filter (HEPA filter)‡ Any continuous flow supplied air respirator equipped with a hood or helmet
≤ 50 µg/m^3 (50 × REL)	Any air-purifying full-facepiece respirator equipped with N-100, R-100, or P-100 filter Any powered air-purifying respirator equipped with a tight-fitting half-facepiece and a high-efficiency particulate air filter. Any negative pressure (demand) supplied-air respirator equipped with a full-facepiece Any continuous flow supplied-air respirator with a tight-fitting half-facepiece Any negative pressure (demand) self- contained respirator equipped with a full-facepiece
≤ 1000 µg/m^3 (1,000 × REL)	Any pressure-demand supplied-air respirator equipped with a full-facepiece

* The protection offered by a given respirator is contingent upon (1) the respirator user adhering to complete program requirements (such as those required by OSHA in 29 CFR 1910.134), (2) the use of NIOSH-certified respirators in their approved configuration, and (3) individual fit testing to rule out those respirators that cannot achieve a good fit on individual workers.

† The appropriate type of particulate filter means: Any 95 or 100 series (N, R, or P) filter. **Note:** N-95 or N-100 series filters should not be used in environments where there is potential for exposure to oil mists.

‡ Some powered air purifying respirators with a hood/helmet are considered to have an APF of 1000 and thus could be used in situations involving higher airborne concentrations of CNT and CNF (< 1000 µg/m3). Contact the respirator manufacturer to determine whether this would apply. Absent

such a determination, powered air purifying respirators with helmets/hoods are to be treated as loose-fitting facepiece respirators, and receive an APF of 25.

Note: complete information on the selection of respirators can be found at (1) OSHA 3352-02 2009, *Assigned Protection Factors for the Revised Respiratory Protection Standard* at http://www.osha.gov/Publications/3352-APF-respirators.html, and (2) NIOSH at [http://www.cdc.gov/niosh/docs/2005-100/default.html].

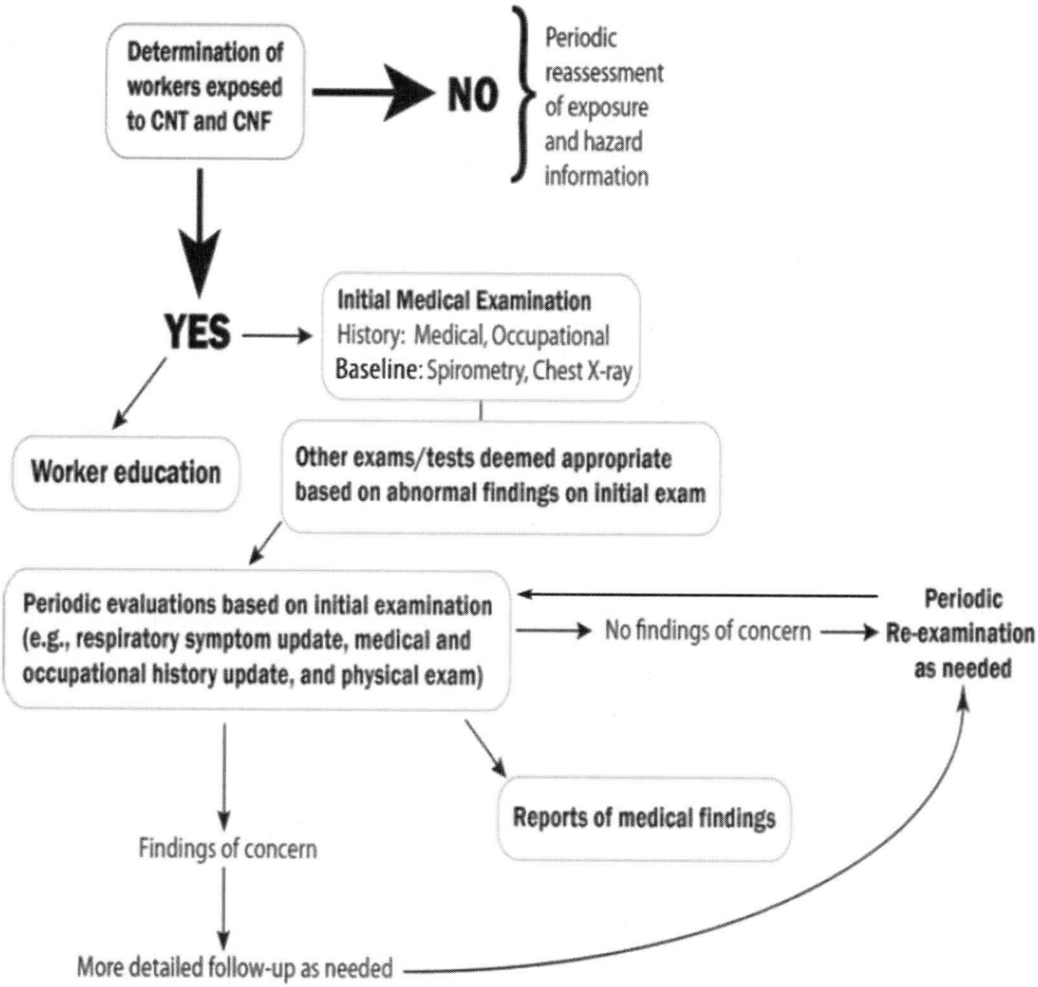

Figure 6–2. Medical screening and surveillance recommendations.

6.7.2. Program Oversight

Oversight of the medical surveillance program should be assigned to a qualified healthcare professional who is informed and knowledgeable about potential workplace exposures, routes of exposure, and potential health effects related to CNT and CNF.

6.7.3. Screening Elements

Initial Evaluation

- An initial (baseline) evaluation should be conducted by a qualified health professional and should consist of the following:
  - An occupational and medical history with respiratory symptoms assessed by use of a standardized questionnaire such as the American Thoracic Society Respiratory Questionnaire [Ferris 1978], or the most recent equivalent.
  - A physical examination with an emphasis on the respiratory system.
  - A spirometry test. (Anyone administering spirometry testing as part of the medical screening program should have completed a NIOSH-approved training course in spirometry or other equivalent training; additionally, the health professional overseeing the screening and surveillance program should be expert in the interpretation of spirometry testing results, enabling follow-up evaluation as needed.)
  - A baseline chest X-ray (digital or film-screen radiograph). All baseline chest images should be clinically interpreted by a board eligible/certified radiologist or other physician with appropriate expertise, such as a board eligible/certified pulmonologist. Periodic follow up chest X-rays may be considered, but there is currently insufficient evidence to evaluate effectiveness. However, if periodic follow up is obtained, clinical interpretation and classification of the images by a NIOSH-certified B reader using the standard International Classification of Radiographs of Pneumoconioses (ILO 2011 or the most recent equivalent) are recommended.
  - Other examinations or medical tests deemed appropriate by the responsible healthcare professional. (The need for specific medical tests may be based on factors such as abnormal findings on initial examination—for example, the findings of an unexplained abnormality on a chest X-ray should prompt further evaluation that might include the use of high-resolution computed tomography scan of the thorax.)

Periodic Evaluations

- Evaluations should be conducted at regular intervals and at other times (e.g., post-incident) as deemed appropriate for the individual worker by the responsible healthcare professional. Evaluations should be based on data gathered in the initial evaluation, ongoing work history, changes in symptoms such as new or worsening respiratory symptoms, and when process changes occur in the workplace (e.g., a change in how CNT or CNF are manufactured or used or an unintentional spill). Evaluations should include the following:
  - An occupational and medical history update, including a respiratory symptom update, and focused physical examination— performed annually.
  - Spirometry testing less frequently than every 3 years is not recommended [OSHA NIOSH 2011]
  - Consideration of specific medical tests (e.g., chest X-ray).

Written reports of medical findings

- The healthcare professional should give each *worker* a written report containing the following:
  - The individual worker's medical examination results.
  - Medical opinions and/or recommendations concerning any relationships between the individual worker's medical conditions and occupational exposures, any special instructions on the individual's exposures and/or use of personal protective equipment, and any further evaluation or treatment.
- For each examined employee, the health-care professional should give the *employer* a written report specifying the following:
  - Any work or exposure restrictions based on the results of medical evaluations.
  - Any recommendations concerning use of personal protective equipment.
  - A medical opinion as to whether any of the worker's medical conditions is likely to have been caused or aggravated by occupational exposures.
- Findings from the medical evaluations having no bearing on the worker's ability to work with CNT and CNF should not be included in any reports to employers. Confidentiality of the worker's medical records should be enforced in accordance with all applicable regulations and guidelines.

6.7.4. Worker Education

Workers should be provided information sufficient to allow them to understand the nature of potential workplace exposures, routes of exposure, and instructions for reporting health symptoms. Workers should also be given information about the purposes of medical screening, the health benefits of the program, and the procedures involved.

6.7.5. Periodic Evaluation of Data and Surveillance Program

Standardized medical screening data should be periodically aggregated and evaluated to identify patterns of worker health that may be linked to work activities and practices that require additional primary prevention efforts [i.e., medical surveillance]. This analysis should be performed by a qualified health-care professional or other knowledgeable person to identify patterns of worker health that may be linked to work activities or exposures. Confidentiality of worker's medical records should be enforced in accordance with all applicable regulations and guidelines.

Employers should periodically evaluate the elements of the medical screening program to ensure that the program is consistent with current knowledge related to exposures and health effects associated with occupational exposure to CNT and CNF.

Other important components related to occupational health surveillance programs, including medical surveillance and screening, are discussed in Appendix B.

7. Research Needs

Additional data and information are needed to assist NIOSH in evaluating the occupational safety and health concerns of working with CNT and CNF. Data are particularly needed on workplace exposures to CNT and CNF, as well as information on whether in-place exposure control measures (e.g., engineering controls) and work practices are effective in reducing worker exposures. Additional assessment of NIOSH Method 5040 is needed to better understand potential interferences or other method limitations, improve the sensitivity and precision of the analytical method, and establish validity through the use of reference materials. The conduct of experimental animal studies with various types of CNT and CNF would help to explain potential mechanisms of toxicity and would provide a better understanding of the exposure parameters (e.g., mass, fiber/structure number, and particle size) that best describe the toxicological responses. Chronic studies in animals are needed to better estimate the longterm risks of lung disease in workers.

The following types of information and research are needed:

7.1. Workplace Exposures, Measurement, and Controls

- Quantify worker airborne exposures to CNT and CNF.
- Evaluate NIOSH Method 5040 and other appropriate sampling and analytical methods in CNT and CNF workplaces. For example, validate Method 5040 against EC reference material and ruggedize against several CNT and CNF types.
- Improve the sensitivity and precision of NIOSH Method 5040 and other appropriate methods for measuring airborne concentrations of CNT and CNF, including those based on metrics that may be more closely associated with the potential adverse effects (e.g., electron microscopy-based CNT or CNF structure counts).
- Develop improved sampling and analytical methods for measuring airborne exposures to CNT and CNF. Apply these different methods in toxicological studies to determine which exposure metric best predicts the health endpoints in laboratory animal studies.
- Determine the effectiveness of engineering controls to control airborne exposures to CNT and CNF below the NIOSH REL of 1 $\mu g/m^3$.
- Confirm the effectiveness of using HEPA filters in an exhaust ventilation system for removing exposures to CNT and CNF.
- Determine the effectiveness of gloves and other PPE barrier materials in preventing dermal exposure to CNT and CNF.
- Identify, quantify, and develop CNT and CNF reference materials for toxicology studies and for measurement quality control.
- Conduct workplace studies to measure total inward leakage (TIL) of respirators for workers exposed to nanoparticles (e.g., CNT/CNF).

7.2. Experimental and Human Studies

- Conduct chronic animal inhalation studies to assess respiratory and other organ (e.g., heart and other circulatory system) effects. Special emphasis should be placed on assessing the risk for developing lung fibrosis and cancer. Studies should evaluate different types of CNT and CNF and use various exposure metrics (e.g., mass, tube, and structure counts, surface area) for assessing toxicological responses.

- Determine the mechanisms and other causative factors (e.g., tube, fiber and agglomerate size, surface area, and surface reactivity) by which CNT and CNF induce adverse effects (e.g., lung fibrosis) in animals.
- Develop early markers of exposure and pulmonary response to CNT and CNF, given evidence from animal studies that CNT and CNF persist in the lungs and result in the development and progression of pulmonary fibrosis and/or cancer at relatively low-mass doses.
- Quantitatively and qualitatively compare the CNT and CNF materials used in the animal studies with the CNT and CNF materials found in workplace air.
- Determine the potential for CNT and CNF to penetrate the skin and cause toxicity.
- Evaluate the predictive value of using *in vitro* screening tests for assessing the hazard (e.g., fibrogenic potential) of various types of CNT and CNF.
- Assess the feasibility of establishing exposure registries for workers potentially exposed to CNT and CNF for conducting future epidemiologic studies and surveillance activities.
- Conduct cross-sectional and prospective studies of workers exposed to CNT and CNF.

APPENDIX A: QUANTITATIVE RISK ASSESSMENT OF CNT

A.1. Introduction

The increasing production and use of CNT and the preliminary significant toxicology findings necessitate an assessment of the potential adverse health effects in workers who produce or use these materials. Risk assessment is a process that uses standardized tools and procedures to characterize the health risk of exposure to a hazardous substance, as well as the uncertainties associated with those risk estimates. Research studies in toxicology, epidemiology, exposure measurement, and other areas provide the data needed to perform the risk assessment. The standard risk assessment paradigm in the United States includes four basic steps: hazard assessment, exposure assessment, dose-response analysis, and risk characterization [NRC 1983, 2009]. Risk assessment also involves the initial steps in problem formulation and evaluation of the various risk management options [NRC 2009]. The most recent guidance [NRC 2009] recommends asking these questions: "What are the options available to reduce the hazards or exposures that have been identified, and how can risk assessment be used to evaluate the merits of the various options?" Risk assessment is intended to provide information needed to determine risk management options.

Risk assessment practice seeks to use the best available data and scientific methods as the basis for public and occupational health decision-making [NRC 2009]. When sufficient dose-response data are available (e.g., from animal studies), quantitative risk assessment can be performed. Quantitative risk assessment provides estimates of the severity and likelihood of an adverse response associated with exposure to a hazardous agent [Piegorsch and Bailer 2005; NRC 2009]. Risk assessments are used in developing occupational exposure limits and in selecting and evaluating the effectiveness of exposure controls and other risk management strategies to protect workers' health.

The best data available for risk assessment, in the absence of epidemiological studies of workers producing or using CNT, are from animal studies with CNT. These studies include two subchronic inhalation studies of MWCNT in rats and several shortterm studies of SWCNT, MWCNT, or CNF in rats or mice. These studies provide the data and information on the dose-response relationships and the biological mechanisms of early-stage inflammatory and fibrotic lung effects from exposure to CNT. No chronic animal studies of CNT were available for this risk assessment.

The biological mode of action for CNT and CNF, as for inhaled particles and fibers, generally relates to their physical and chemical properties. These properties include: (1) nano-structure which increases the surface area and associated inflammogenic and fibrogenic response; (2) fiber shape which may decrease clearance of long structures, resulting in translocation to the interstitial and pleural tissues of the lungs; and (3) the graphitic structure of CNT and CNF which influences their durability and biopersistence [Donaldson et al. 2006; Shvedova et al. 2009; Castranova 2011]. CNT and CNF are heterogeneous structures, and differences can include size (length and diameter), metal contaminants (type and amount), surface chemistry, and tendency to aggregate/agglomerate. CNT and CNF typically form agglomerates in air but may also exist as individual structures Johnson et al. 2010, Methner et al. 2010, Dahm et al. 2011]. Evidence from shortterm and subchronic studies in animals indicates that CNT may be biopersistent in the lungs [Muller et al. 2005; Deng et al. 2007; Elgrabli et al. 2008b; Mercer et al. 2009; Pauluhn 2010a,b] However, some evidence suggests that functionalization may increase biodegradation of CNT [Kagan et al. 2010; Osmond-McLeod et al. 2011].

Dose metrics that have been associated with lung responses to CNT or CNF in animal studies include mass, volume, number, and surface area. The CNT volume dose was associated with the overloading of CNT clearance from rat lungs and to the lung responses [Pauluhn 2010a]. The specific surface area (m^2/g) dose of various types of CNTs was associated with the pulmonary inflammation response in rats [Nakanishi 2011a]. Mercer et al. [2011] showed that on a mass basis, SWCNT is more fibrogenic than MWCNT, but this difference disappeared after accounting for the greater specific surface area of the SWCNT than MWCNT. Murray et al. [2012] found that the effective surface area (estimated from the geometry of the structures observed by electron microscopy) was more closely associated with the pulmonary inflammation and fibrosis in mice.

In addition to exhibiting some of the same physical-chemical properties of other poorly-soluble particles and/or fibers, the nanoscale structure of CNT and CNF may relate to more specific biological modes (or mechanisms) of action. For example, evidence *in vitro* suggests that disperse CNT may act as a basement membrane, which enhances fibroblast proliferation and collagen production [Wang et al. 2010]. This mechanism is consistent with the observation in mice of the rapid onset of diffuse interstitial fibrosis, which progressed in the absence of persistent inflammation, following exposure to SWCNT or MWCNT by pharyngeal aspiration [Shvedova et al. 2005, 2008; Porter et al. 2010]. As fibrosis progresses, it causes thickening of the alveolar septal air/blood barrier, which can result in a decrease of gas-exchange between lung and blood [Hubbs et al. 2011].

The focus of this quantitative risk assessment is on the earlystage noncancer lung responses (fibrotic and inflammatory) from studies in rats and mice, for which dose-response data are available. These responses are relevant to humans as observed in workers in dusty jobs [Rom and Markowitz 2006; Hubbs et al. 2011]. Dose-response relationships are based on mass dose, because the mass of CNT and CNF was associated with lung responses in all of the animal studies and because it is the metric typically used to measure airborne exposure in the workplace (Section 6 and Appendix C). The evidence for cancer effects from CNT and CNF (Section 3 and 4) is insufficient for quantitative risk assessment, and may also depend on specific types of CNT or CNF structures [Donaldson et al. 2011; Nagai et al. 2011; Schulte et al. 2012].

A.2. Methods

NIOSH used dose-response data from subchronic and shortterm studies in rats and mice exposed to SWCNT or MWCNT to estimate the lung doses associated with earlystage inflammatory and fibrotic lung responses. Benchmark dose (BMD) modeling [Crump 1984; 1995; US EPA 2010] of rodent dose-response data was used to estimate an adverse effect level (10% excess risk of earlystage lung effects). The animal BMD was extrapolated to humans to estimate the risk of working lifetime exposures and provided the scientific basis for developing a NIOSH recommended exposure limit (REL) for CNT. Dose-response data from subchronic and shortterm studies in rats and mice exposed to SWCNT or MWCNT were used to estimate the BMDs associated with benchmark responses (BMRs) of earlystage inflammatory and fibrotic lung responses. The rodent-based BMD estimates were extrapolated to humans by accounting for species differences in factors influencing lung dose in order to estimate the working lifetime risk of airborne exposure to CNT. Sensitivity analyses were performed to evaluate the influence of the various methods and assumptions on the risk estimates.

When feasible, NIOSH utilizes BMD estimates in risk assessment rather than a lowest observed adverse effect level (LOAEL) or a no observed adverse effect level (NOAEL) for the following reasons: (1) BMD methods provide a standardized approach for risk estimation; (2) BMD methods provide both maximum likelihood (BMD) and 95% lower confidence limit, or BMDL, estimates, which explicitly account for the sample size and variability in the data; and (3) BMD models efficiently use all of the dose-response data in estimating the BMD(L)s. In contrast, NOAEL and LOAEL estimates are: (1) not risk-based; (2) generally interpreted as estimates of a threshold; and (3) sensitive to the study size and dose group spacing. However, BMD estimation may require more dose-response data than does a NOAEL or LOAEL. Sparse data provide limited information for BMD estimation and can result in model uncertainty. In addition, the NOAEL estimate may depend on the size of the study, such that larger studies (e.g., greater number of animals) could detect effects at smaller exposures. This would not occur with the BMDL estimate, which would "simply approach the (true) BMD" with increasing study size [Crump 2002]. Comparisons of the BMD(L) estimates to the LOAELs or NOAELs provide an informal check on the estimated and observed responses in the low-dose region of the data; and sensitivity analyses of the methods and assumptions in these analyses provides information on the qualitative and quantitative uncertainty in these risk estimates and derived OELs (Section A.6).

A.2.1. Rodent Dose-response Data

A.2.1.1. Data Selection

The published rodent studies on pulmonary responses to CNT (Section 3, Tables 1–3c) were examined for possible inclusion in this risk assessment. Pulmonary effects were examined because of their relevance to workers who may be exposed to CNT in workplace air. The studies with adequate quantitative dose-response data to estimate BMDs were included in these analyses. These studies reported on the size and characterization of the CNT (Table A–1) as well as the route of exposure, doses, duration of exposure or post-exposure, number of animals per group, and lung responses. In general, the CNT animal studies have limited data, with few (4–20) animals per dose group and sparse dose group spacing, especially in the low range of the dose-response curve. Some of these studies just meet the minimum data criteria for BMD estimation, that is, a significant dose-related trend in the selected endpoint [US EPA 2012]. It is preferable to have data with one or more doses near the benchmark response (e.g., 10%) [US EPA 2012]; however, in some studies the response proportions were quite high at each dose (e.g., 30–100%) [Lam et al. 2004; Ma-Hock et al. 2009; Pauluhn 2010a]. In addition, one study [Shvedova et al. 2008] had only one dose group in addition to the control, but the study was included because it is the only animal inhalation study for SWCNT currently available and it provides a useful comparison by route of exposure. No other deficiencies were noted in the selected studies that would have resulted in their omission.

Either the individual animal dose-response data or the mean and standard deviation of the group response are required for BMD model fitting. The dose was either the intratracheal instillation (IT) or pharyngeal aspiration (PA) administered mass dose (mg/lung) or the inhaled mass concentration (mg/m^3). Datasets with treatment-related mortality of animals were not used. Data on special preparations of CNT (e.g., ground CNT) or studies using sensitive animal models (e.g., vitamin E deficient) were not included (although these data may be of interest for subsequent analyses using animal models to investigate biological mechanisms, including in sensitive human populations, or to evaluate the effect of specific alterations in CNT properties on hazard potential).

Study details of the data selected for this risk assessment are provided in Table A–1. These studies include the two recently published subchronic inhalation studies of MWCNT in rats [Ma-Hock et al. 2009; Pauluhn 2010a] and several IT, PA, or shortterm inhalation studies in rats or mice exposed to SWCNT [Lam et al. 2004; Shvedova et al. 2005, 2008] or MWCNT [Muller et al. 2005; Mercer et al. 2011] with post-exposure durations and examination from 4 to 26 weeks after exposure. In the subchronic inhalation studies, rats were head-nose exposed [Ma-Hock et al. 2009] or nose-only exposed [Pauluhn 2010a] to three or four different airborne mass concentrations (6 hr/d, 5 d/week) for 13 weeks. Lung responses were examined at the end of the 13-week exposure in both studies; post-exposure follow-up was extended to 6 months in the Pauluhn [2010a] study.

The IT, PA, and shortterm inhalation studies provide additional dose-response data for comparison to other MWCNT or SWCNT with different types and amounts of metal contaminants. Although both IT and PA routes bypass the head region and deliver the CNT material directly to the trachea and lung airways, PA is to be considered more similar to inhalation than IT because PA provides greater dispersion of deposited material in the lungs [Shvedova et al. 2005, 2008]. Following the administered dose (on day 1), the lung responses were evaluated after a post-exposure period (e.g., 1, 7, 28, 60, and/ or 90 days). For studies with more than one post-exposure duration, the longest post-exposure duration data are used in these risk analyses. Some of these studies also provide dose-response data on other particles or fibers (e.g., ultrafine carbon black, crystalline silica, and asbestos) for comparison of dose and response to that observed from exposure to MWCNT or SWCNT [Lam et al. 2004; Muller et al. 2005; Shvedova et al. 2005]. Two other shortterm exposure studies (Porter et al. [2010] and Ellinger-Ziegelbauer-Pauluhn [2009]), which were included in the external review draft of the CIB [NIOSH 2010], have been omitted in this analysis This is because the dose-response data were of equivocal fit to the minimum data criteria for BMD analysis and because updates of these studies are available for the same CNT material and from the same laboratory (i.e., Mercer et al. [2011] and Pauluhn [2010a]). Those studies are included in these risk analyses.

A.2.1.2. Dose Rate Evaluation

A study of 1-day inhalation exposure to MWCNT (Baytubes) in rats and examined 13 weeks after the end of exposure [Ellinger-Ziegelbauer and Pauluhn 2009] provided an opportunity to compare the dose-response relationships of the 1-day inhalation exposure study with that of the 13-week (subchronic) inhalation study [Pauluhn 2010a] in order to examine the influence of dose rate on the rat lung responses (Section A.3.2). These findings are relevant to interpreting and using the results from the shortterm exposure studies of the SWCNT and other MWCNT.

A.2.1.3. Lung Responses Evaluated

The lungs are the target organ for adverse effects as shown in animal studies of CNT (Sections 3 and 4). Granulomatous inflammation, alveolar interstitial thickening, and pulmonary fibrosis are among the benchmark responses evaluated in this risk assessment (Table A–1). These responses are considered to be relevant to workers since inflammatory and fibrotic lung diseases have been associated with occupational exposure to various types of inhaled particles and fibers [Rom and Markowitz 2006]. These pulmonary inflammation and fibrotic effects in animals were observed at relatively early stages, although they developed earlier in mice exposed to SWCNT than from exposure to crystalline silica, which is a known fibrogenic particle [Shvedova et al. 2005; Lam et al. 2004].

The most quantitative measure of fibrosis was reported by the studies that measured the thickening of the gas-exchange region of the lungs (alveolar interstitial or septal connective tissue) due to increased collagen (as observed by lung tissue staining in histopathology examination) [Mercer et al. 2010, 2011; Shvedova et al. 2005, 2008; Hubbs et al. 2011]. This alveolar thickening was observed to progress with time after administration of a single dose in mice administered by PA [Shvedova et al. 2005; Mercer et al. 2008; Porter et al. 2010; Mercer et al. 2011]. Alveolar thickening was also observed in a subchronic study, which persisted up to 6 months after the end of exposure in a 13-wk inhalation study in rats [Pauluhn 2010a]. Alveolar interstitial (epithelial cell) thickness has been used as the adverse response in other risk assessment (of ozone) because it indicates "fundamental structural remodeling" [US EPA 1996; Stockstill et al. 1995].

Alveolar interstitial fibrosis can be detected by Sirius red staining of septal collagen [Hubbs et al. 2011]. Interstitial thickening with fibrosis has been demonstrated by Sirius red staining of lungs from mice exposed to SWCNT or MWCNT [Shvedova et al. 2005, 2008;

Mercer et al. 2011]. In SWCNT exposed mice, the septal fibrosis has been further confirmed by transmission electron microscopy [Mercer et al. 2008]. Pauluhn [2010a] also reported alveolar interstitial thickening in rats exposed to MWCNT, but distinguished the focal effects observed at 0.4 mg/m^3 from those at higher exposures. That is, Pauluhn [2010a] reported: "Increased interstitial collagen staining (Sirius red) occurred at 1.5 and 6 mg/m^3. Focal areas of increased collagen staining were adjacent to sites of increased particle deposition and inflammatory infiltrates (onset at 0.4 mg/m^3, see Table 3). Increased septal collagen staining was depicted as equal to interstitial fibrosis (for details, see Figure 12)." In that study, a severity level of minimal (category 1) or greater persisted or progressed up to 26 weeks after the end of the 13-week inhalation exposure to either 0.4, 1.5, or 6 mg/m^3 [Pauluhn 2010a, Table 3]. Hypercellularity in the bronchial alveolar junctions was observed in these same dose groups; this effect persisted after the end of exposure, but resolved by the 39th week in the 0.4 mg/m^3 group. The 0.4 mg/m^3 dose group was considered the LOAEL for inflammatory lung effects, while 0.1 mg/m^3 was considered the NOAEL [Pauluhn 2010a]. Concerning the focal septal thickening observed at 0.4 mg/m^3, pathologists' interpretations may differ as to whether these earlystage responses would be considered adverse or to have the potential to become adverse. NIOSH interpreted the alveolar septal thickening (and associated effects) in the 0.4 mg/m^3 and higher dose groups as being adverse changes of relevance to human health risk assessment due to their persistence and consistency with the earlystage changes in the development of pulmonary fibrosis. For these reasons, the alveolar septal thickening of minimal or higher grade (i.e., proportion of rats with this response, which included rats exposed at 0.4 mg/m^3 and higher doses) was selected as the benchmark response in the Pauluhn [2010a] study. Although these data were reported as the average histopathology score in each dose group [Pauluhn 2010a, Table 3], NIOSH requested the response proportion data as these were needed for the dichotomous BMD modeling. These data were provided by Dr. Pauluhn in response to this request [personal communication, J. Pauluhn and E. Kuempel, 1/27/10].

Pulmonary inflammation has been associated with exposure to airborne particles and fibers, and it is a hallmark of occupational lung disease in humans. It is also a precursor to particle-associated lung cancer in rats [IARC 2010; NIOSH 2011a]. Pulmonary inflammation can be measured by the increase in polymorphonuclear leukocytes (PMNs) in bronchoalveolar lavage (BAL) fluid following exposure to various particles including CNT. However, for some CNT, the inflammation resolves, while the fibrosis continues to develop [Shvedova et al. 2005, 2008; Mercer et al. 2010; Pauluhn 2010a]. This indicates that neutrophilic inflammation in BAL fluid may not be a good predictor of adverse lung effects from some CNT, which appear to cause fibrosis by a different mechanism than for other types of particles and fibers (by resembling the lung basement membrane and serving as a framework for fibroblast cell growth, without eliciting a persistent inflammatory response) [Wang et al. 2010b]. In other studies, the inflammatory effects of MWCNT were associated with granuloma development [Ma-Hock et al. 2009] and with alveolar lipoproteinosis, a more severe inflammatory lung response, observed at higher doses of MWCNT [Ma-Hock et al. 2009].

Minimal or higher levels of severity of these lung responses were selected as the benchmark responses. This included minimal level (grade 1 or higher) of pulmonary inflammation [Ma-Hock et al. 2009] or alveolar septal thickening [Pauluhn 2010a] as observed by histopathology. The incidence data on the minimal level of effect that is persistent provides a sensitive measure of a critical effect, which is of interest for health risk assessment. It is not known whether the human-equivalent effects to those observed in the animal studies would be associated with abnormal lung function or clinical disease, or if progression to more severe levels could occur if these effects developed as a result of chronic exposure. To evaluate sensitivity of risk estimates to the selection of a minimal level of disease, risk estimates were also derived for the next level of response (grade 2 or higher) in the subchronic animal studies.

The lung response measures in this risk assessment are either dichotomous (proportion of animals observed with the response endpoint) or continuous (amount or level of response in individual animals) (Table A–1). The dichotomous responses include the incidence of lung granulomas [Lam et al. 2004]; granulomatous inflammation [Ma-Hock et al. 2009]; and histopathology grade of alveolar interstitial (septal) thickening [Pauluhn 2010a]. The continuous responses include the amount of hydroxyproline (as mass) [Muller et al. 2005] and alveolar interstitial connective tissue thickness [Shvedova et al. 2005, 2008; Mercer et al. 2011].

A.2.1.4. Summary of Dose-response Data

Collectively, the data available for CNT risk assessment include dose-response data from several rodent species and strains, both males and females, and three routes of exposure to several types of SWCNT and MWCNT with varying types and amounts of metal contaminants (Table A–1). The dose metric used in this risk assessment is the mass dose of CNT in the lungs, either the administered dose (IT or PA studies) or the lung burden (deposited or retained) estimated from the airborne particle size distribution and exposure concentration data (inhalation studies). Mass dose was used because all of the studies reported this dose metric and because mass dose was associated with the inflammatory and fibrotic lung responses in the animal studies.

A.2.2. Estimated Lung Dose in Animals

For the IT and PA studies [Lam et al. 2004; Shvedova et al. 2005; Muller et al. 2005], the administered CNT mass dose was assumed to be equivalent to the deposited lung dose. In the inhalation studies [Shvedova et al. 2008; Ma-Hock et al. 2009; Ellinger-Ziegelbauer and Pauluhn 2009; Pauluhn 2010a], the deposited lung dose was estimated from the exposure concentration and duration, the species-specific ventilation rate, and the alveolar deposition fraction (estimated from the CNT aerodynamic particle size data), as follows:

Equation A–1:

$$\text{Deposited lung dose } (\mu g) = \text{Exposure Concentration } (\mu g/m^3) \times \text{Duration } (hr/d \times d/wk \times wk) \times \text{Minute Ventilation } (L/min) \times 0.001\ m^3/L \times 60\ min/hr \times \text{Alveolar Deposition Fraction}$$

The exposure concentration and duration, as reported in the animal studies, are shown in Table A–1. The values used for respiratory minute ventilation were based on the species and body weight: 0.037 L/min for mice [US EPA 1988; 2006]; 0.25 L/ min for male rats in Pauluhn [2010] (369 g body weight); and 0.21 L/min for male and female rats in MaHock et al. [2009], assuming average body weight of 300 g [US EPA 1994, 2006]. The alveolar lung deposition fraction in rats was estimated from the MPPD 2.0 model for inhaled poorly soluble spherical particles [CIIT and RIVM 2006] using the mass-median aerodynamic diameter (MMAD) and geometric standard deviation (GSD) data reported for SWCNT and MWCNT (Table A–2). In the mouse inhalation study [Shvedova et al. 2008], an alveolar deposition fraction of 0.01 was estimated based on the MMAD (Table A–2) and interpolating from the deposition fractions for monodisperse spherical particles reported in Table 2 of Raabe et al. [1988]. For the two subchronic inhalation studies of MWCNT [Ma-Hock et al. 2009; Pauluhn 2010a], in addition to the deposited dose, the retained lung dose was also estimated. The MPPD 2.0 model [CIIT and RIVM 2006] was used to estimate the lung burden at the end of the 13-week exposure based on the particle MMAD and GSD (Table A–2) reported in those studies, assuming unit density (the lowest density accepted by MPPD 2.0).

Table A–1. Rodent study information

Rodent study	CNT type and main metal component	Species, strain, gender	Route of exposure*	Number of animals per dose group	Dose (& exposure duration, if inhalation)	Post-exposure days	Lung response
Lam et al. [2004]	SWCNT Fe 2.0%	Mouse, B6C3F1, male	IT	5	0, 0.1, or 0.5 mg	90	Granuloma†
Shvedova et al. [2005]	SWCNT Fe 0.2%	Mouse, C57BL/6, female	PA	6 (28 d) 3 (60 d)	0, 10, 20, 40 μg	1, 3, 7, 28, 60	Alveolar connective tissue thickness‡
Muller et al. [2005]	MWCNT Al 2%, Co 0.5%, Fe 0.5%	Rat, Sprague-Dawley, female	IT	5	0, 0.5, 2, 5 mg	28 and 60	Hydroxyprolin e amount‡
Shvedova et al. [2008]	SWCNT Fe 18%	Mouse, C57BL/6, female	Inhal	5	5 mg/m^3 (5 hr/d, 4 d)	1, 7, 28 (after 4 d exposure)	Alveolar connective tissue thickness‡
Ma-Hock et al. [2009]	MWCNT Al_2O_3 9.6%	Rat, Wistar (Crl:WI), male and female	Inhal	20 (10 each gender)	0, 0.1, 0.5, 2.5 mg/m^3 (6 hr/d, 5 d/wk, 13 wk)	1	Granulomatous inflammation (minimal or greater)†
Pauluhn [2010]	MWCNT Co 0.5%	Rat, Wistar (HsdCpb: WU), male	Inhal	10	0, 0.10, 0.45, 1.62, 5.98 mg/m^3 (6 hr/d, 5 d/wk, 13 wk)	1, 28, 91, 182	Alveolar septal thickening (minimal or greater)†
Mercer et al. [2011]	MWCNT Fe 0.3%	Mouse, C57BL/6J, male	PA	6	0, 10, 20, 40, 80 μg	1, 7, 28, 56	Alveolar connective tissue thickness‡

*Intratracheal instillation (IT); pharyngeal aspiration (PA); inhalation (inhal).

†Dichotomous response.

‡Continuous response.

However, Ma-Hock et al. [2009] reported the MWCNT particle density of 0.043 g/ ml, and Pauluhn [2010a] reported the MWCNT particle density of 0.1–0.3 g/ml. The sensitivity of the lung dose estimates to the assumption of density 1 or lower is evaluated in Section

A.6.1.1. Also evaluated is the effect of MPPD model version 2.0 or 2.1 on the lung dose estimates. A recent update of MPPD 2.0 [CIIT and RIVM 2007] to MPPD 2.1 [ARA 2011] included revised estimates of the rat head/extra thoracic deposition efficiency based on the equations in Raabe et al. [1988], which resulted in lower predicted deposition fractions in the rat pulmonary region [personal communication, O. Price and E. Kuempel, 9/24/10] (Section A.6.1.1).

These lung dosimetry models have not been evaluated for CNT. However, the measured aerodynamic diameter is considered to provide a reasonable estimate of the deposition efficiency in the respiratory tract for CNT which have MMAD in the micrometer size range (Table A–2) (see Section A.6.1.1 for further discussion). The estimates of CNT clearance and retention in the lungs may be more uncertain than those for deposition fraction, given the slower clearance reported for CNT in some animal studies [Mercer et al. 2009; Pauluhn 2010a,b]. Reasonable bounds on the uncertainty of the CNT lung dose estimates are considered to be the deposited (no clearance) and the retained (normal clearance) doses predicted from the spherical-particle based models [MPPD, CIIT and RIVM 2007, ARA 2011]. This is because the CNT deposited in the lungs may undergo some clearance, although evidence from animal studies suggests the clearance rate may be slower than for other poorly soluble particles at relatively low-mass doses in the rats and mice [Deng et al. 2007; Mercer et al. 2009; Pauluhn 2010a; 2011; Elgrabli et al. 2008b].

A.2.3. Animal Dose-response Modeling and BMD Estimation

The dose-response data in rats and mice exposed to SWCNT or MWCNT were modeled using benchmark dose methods [Crump 1984; US EPA 2010]. A benchmark dose has been defined as ". . . a statistical lower confidence limit for the dose corresponding to a specified increase in level of [adverse] health effect over the background level" [Crump 1984]. The increased level of adverse effect (called a benchmark response, or BMR) associated with a BMD is typically in the low region of the dose-response data (e.g., a 10% excess risk). In this document, the term BMD is used to describe the point estimate based on maximum likelihood estimation, and the term BMDL is used to describe the lower 95% confidence limit (i.e., as originally defined by Crump [1984]). A 10% excess risk, based on dichotomous or quantal data, is used because it is at or near the limit of sensitivity in the animal bioassay [US EPA 2012; Crump 2002]. The BMDL associated with a 10% BMR is used as a point of departure (POD) for low-dose extrapolation using linear or nonlinear methods (depending on the mode of action evidence) [US EPA 2012]. The low-dose extrapolation may include estimation of the probability of effects at low doses or below a reference value (not risk-based) by accounting for uncertainties in the dose estimation (e.g., extrapolation from animal to human, inter-individual variability, limitations in the animal data [US EPA 2012].

A.2.3.1. Dichotomous Response Data

For dichotomous data (yes/no response), a BMD is defined as the dose associated with a specified increase in the probability of a given response, either as an excess risk (i.e., additional probability above background) or as a relative risk (i.e., relative to the background probability of having a normal response) [Crump 2002].

In this analysis, the BMD (using dichotomous data) is the dose d corresponding to a specified excess (added) risk (e.g., 10%) in the proportion of animals with a given adverse lung response (BMR), as follows:

Equation A–2:

$$BMR = P(d) - P(0)$$

where P(d) is the probability of an adverse response at the BMD, and P(0) is the probability of that adverse response in an unexposed population [Crump 2002; US EPA 2010].

The dichotomous BMR lung responses include the presence or absence of granulomatous inflammation [Ma-Hock et al. 2009] or alveolar septal thickening [Pauluhn 2010a] (Table A–1). The propor- tion of animals responding with the minimal or higher severity was selected as the benchmark response. The BMD(L) estimates are expressed as the mass dose of SWCNT or MWCNT in rodent lungs associated with the specified BMR. These animal-based BMD(L)s are extrapolated to humans based on species-specific differences in the estimated deposition and retention of CNTs in the lungs (Section A.2.4).

A.2.3.2. Continuous Response Data

BMD estimation using continuous data requires specifying a BMR level along the continuum of responses. Continuous response data provide information on the amount or degree of a biological response. Continuous response measures may include nonzero levels that are associated with the normal structure or function (e.g., a certain number immune cells or amount of protein in healthy lungs). These levels can become elevated in response to a toxicant, and at some point, they may result in irreversible, functional impairment of the lungs [NIOSH 1986]. If data are available, the BMRs can be based on a biologically significant response that is associated with, or expected to result in, a material impairment of health. However, there may be insufficient data to determine a specific level that is associated with a measurable adverse response. In that case, a statistical criterion may be used as a BMR for continuous data [Crump 1995].

A statistical method (originally referred to as a "hybrid" method) is described by Crump [1995] to provide BMD(L) estimates from continuous data that are equivalent to a 10% excess risk based on dichotomous data, assuming that an abnormal or biologically significant response is defined as the upper 99th percentile of the unexposed (control) distribution. According to this method, "for a normal distribution with constant variance, setting BMR = 0.1 and $P_O = 0.01$ is equivalent to choosing the BMD to be the dose that results in an increase in the mean equal to 1.1 times the standard deviation," assuming a normal distribution with constant variance [Crump 1995]. That is, if one assumes that the probability of the specified adverse response in the unexposed population is the upper 1% of a normal distribution of responses, then selecting a BMR of 1.1 standard deviations above the control mean response is equivalent to a 10% BMD as estimated in dichotomous data.

In evaluating possible BMRs for the continuous data of CNT in mice, earlier studies of chronic ozone exposure in rats were examined to determine if a biologically-based BMR could be identified for pulmonary fibrosis (measured as alveolar connective tissue thickening) associated with abnormal pulmonary function [Chang et al. 1992; Costa et al. 1995; Stockstill et al. 1995]. However, those rat findings did not appear to extrapolate well to the mice in Shvedova et al. [2005, 2008]. That is, the observed abnormal response in rats (associated with a persistent lung function deficit) was a 36% increase in the control mean alveolar connective tissue thickness [Chang et al. 1992; Costa et al. 1995]; however, this amount of response occurred in up to 30% of the control (unexposed) mice in Shvedova et al. [2005, 2008] (vs.

2.5% of controls in Chang et al. [1992], in part due to the greater variability in the alveolar tissue thickness in the unexposed mice. In addition, no data were found of a biologically relevant BMR for the amount of hydroxyproline in the lungs of rats or mice. In the absence of an identified biological basis for a BMR for the continuous response measures of alveolar connective tissue thickening or the amount of hydroxyproline, NIOSH used the statistical criterion described by Crump [1995], in which a BMR of 1.1 standard deviations above the control mean response is equivalent to a 10% excess risk in the dichotomous data, assuming the 99th percentile of the distribution of control responses is abnormal or biologically significant.

That is, the BMR for the continuous data (alveolar connective tissue thickness and hydroxyproline amount) is defined as follows:

Equation A–3:

$$BMR = \mu(d) - \mu(0)$$

where $\mu(d)$ is the mean response at the BMD (d); $\mu(0)$ is the control mean response; and BMR is the specified number of standard deviations (SDs) (i.e., 1.1 in these analyses). Thus, the continuous data-based BMD is the dose associated with a 10% increase in the proportion of animals exposed at dose d with response greater than the 99th percentile of the control mean response. The estimates of $\mu(d)$and $\mu(0)$ are derived from the fitted dose-response models (polynomial) (Section A.2.3.3).

A.2.3.3. BMD Model Fitting

The animal dose-response data were fit using the benchmark modeling software (BMDS 2.1.2) [US EPA 2010]. The dichotomous data were fit with a multistage (polynomial degree 2) model. This is the only model that provided an adequate fit to the subchronic inhalation data, each of which [Ma-Hock et al. 2009; Pauluhn 2010] had only one dose between zero and 100% response for the endpoints evaluated (granulomatous inflammation or alveolar septal thickening, histopathology grade 1 or higher).

The other BMDS models failed to converge or, in further statistical evaluation, showed non-unique parameter solutions. The continuous dose-response data were fit with a polynomial model of degree 2 for all data with three or more dose groups, and degree 1 (linear) for data with two groups (see Table A–1 for dose groups).

P values for goodness of fit were computed for the individual BMDS models (based on likelihood methods) [US EPA 2007]. Model fit was considered adequate at $P > 0.05$ (i.e., testing for lack of fit), although the *P* values based on likelihood ratio tests may not be a reliable indicator of model fit in the studies with few animals per group. The number of animals per dose group in each study is given in Table A–1. EPA typically uses a $P > 0.1$ criteria for BMD model fit [US EPA 2012]. Either criteria is considered reasonable and represents a trade-off in the type I or type II error. That is, $P > 0.1$ provides more power to reject an incorrect model, while $P > 0.05$ provides less chance of rejecting a correct model. The BMD model fits to each data set are shown in Figure A–1 (subchronic studies), Figure A–2 (shortterm studies, dichotomous response), and Figure A–3 (shortterm studies, continuous response).

Table A–2. CNT particle size and alveolar deposition fraction in rodent and human

Study	Particle size information	Human DF_{alv} * and MMAD (GSD) used	Rodent DF_{alv} (same MMAD and GSD)
Lam et al. [2004]	Not reported; same SWCNT source as Shvedova et al. [2005]. (assume same MMAD [GSD] as Shvedova et al. [2008])	0.076 3.5 (2.14)	ad¶
Shvedova et al. [2005]	1–4 nm width (primary particles) (assume same MMAD [GSD] as Shvedova et al. [2008])	0.076 3.5 (2.14)	ad
Muller et al. [2005]	9.7 nm width; 5.9 µm length (primary particles) (assume same MMAD [GSD] as Ma-Hock et al. [2009])	0.099 1.2 (2.7)	ad
Shvedova et al. [2008]	0.8–1.2 nm width; 100–1000 nm length (primary particles); 4.2 µm mass mode diameter; 240 nm count mode diameter; 3.5 µm MMAD (2.14 GSD)§	0.076 3.5 (2.14)	0.01‡
Ma-Hock et al. [2009]	1.5 (3.6); 1.2 (2.7); 0.8 (2.8) µm MMAD (GSD) at 0.1, 0.5, and 2.5 mg/m^3, respectively; (median of 3 values at each concentration). Primary particles: 5–15 nm width; 0.1–10 µm length	0.099 1.2 (2.7)	0.072§
Pauluhn [2010a]	3.05 (1.98); 2.74 (2.11); 3.42 (2.14) µm MMAD (GSD) at 0.4, 1.5, and 6 mg/m^3, respectively. Primary particles: ~10 nm width; 200–1,000 nm length	0.086 2.74 (2.11)	0.046§
Mercer et al. 2011	1.5 µm MMAD from Porter et al. [2010] (GSD not reported; assume 2); count mean width (49 nm; 13.4 SD); median length 3.86 µm (1.94 GSD)	0.10 1.5 (2)	ad¶

* MPPD 2.0 human; Yeh and Schum deposition model; 9.6 m^3/8 hr d (20 L/min, or 1143 ml tidal volume at 17.5 breaths/min); inhalability adjustment; assumed unit density.

† MMAD and GSD in Shvedova et al. [2008] were estimated from data reported in Baron et al. [2008] [personal communication from B. Chen to E. Kuempel, August 4, 2009].

‡ Raabe et al. [1988]: mouse DF_{alv} interpolated from values in Table 2 of Raabe et al. [1988].

§ MPPD 2.0 rat; 0.21 L/min or 2.45 ml tidal volume (assuming 300 g male and female rats) [Ma-Hock et al. 2009]; and 0.25 L/min or 2.45 ml tidal volume (369 g male rats) [Pauluhn 2010; US EPA 1994; 2006]; inhalability adjustment; assumed unit density.

¶ ad—administered dose by intratracheal instillation or pharyngeal aspiration.

A.2.3.4. Human-equivalent Dose and Working Lifetime Exposure

The rodent BMD(L)s were extrapolated to humans based on species-specific differences in the alveolar epithelial surface area of the lungs (i.e., by normalizing the dose per unit of cell surface area). It is assumed that humans and animals would have equal response to an equivalent dose (i.e., mass of CNT per unit surface area of lungs). The human-equivalent BMD and BMDL estimates were the target lung doses used to estimate, respectively, the maximum likelihood estimate (MLE) and 95% lower confidence limit (95% LCL) estimate of the MLE, as an 8-hr TWA exposure concentration during a 45- year working lifetime.

The human-equivalent BMD and BMDL estimates were calculated as follows:

Equation A–4:

Human-equivalent BMD(L) = Rodent BMD(L) × [AlvSA human / AlvSA rodent]

where the values used for alveolar lung surface area (AlvSA) were 102 m^2 (human) [Stone et al. 1992]; 0.4 m^2 (rat) and 0.055 m^2 (mouse) [Mercer et al. 2008]. In Tables A–3 through A–5, the human-equivalent BMD(L)s were multiplied by 0.001 mg/µg to obtain the units of mg per lung.

The human-equivalent BMD(L)s are expressed as the mass (mg) of CNT in the lungs. The working lifetime airborne mass concentration that would result in the BMD(L) human-equivalent lung mass dose was calculated based on either deposition only (no lung clearance) or retention (lung deposition and clearance), as described below.

(a) Deposited lung dose

Equation A–5:

Estimated 8-hr TWA (µg/m^3) = Human-equivalent BMD(L) (µg) / [8-hr worker air inhaled (m^3/day) × Alveolar Deposition Fraction × Work Days]

The values assumed include 9.6 m^3 8-hr air intake (reference worker [ICRP 1994]); alveolar deposition fraction based on aerodynamic particle size (Table A–2); and working lifetime days (250 days/ yr × 45 yr).

(b) Retained lung dose

The MPPD 2.0 human model [CIIT and RIVM 2006] for inhaled poorly soluble spherical particles was used to estimate the working lifetime exposure concentration that would result in the human-equivalent BMD(L) lung burden estimates. This was done by a systematic search to identify the 8-hr time weighted average (TWA) airborne concentration over a 45-year working lifetime that predicted the target lung burden. The input parameters used in the MPPD human model (Yeh and Schum human deposition model option) include CNT aerodynamic particle size (MMAD, GSD) (Table A–2); inhalability adjustment; oronasal-normal augmenter; and reference worker conditions, including 9.6 m^3 of air inhaled per 8-hr day (corresponding to 17.5 breaths/min and tidal volume of 1143 ml), and work for 8 hr/d, 5 d/wk, 50 wk/yr, for 45 years.

In the two subchronic inhalation studies for MWCNT, excess risk estimates were derived based on either the estimated deposited lung dose or the estimated retained lung dose [Ma-Hock et al. 2009; Pauluhn 2010a].

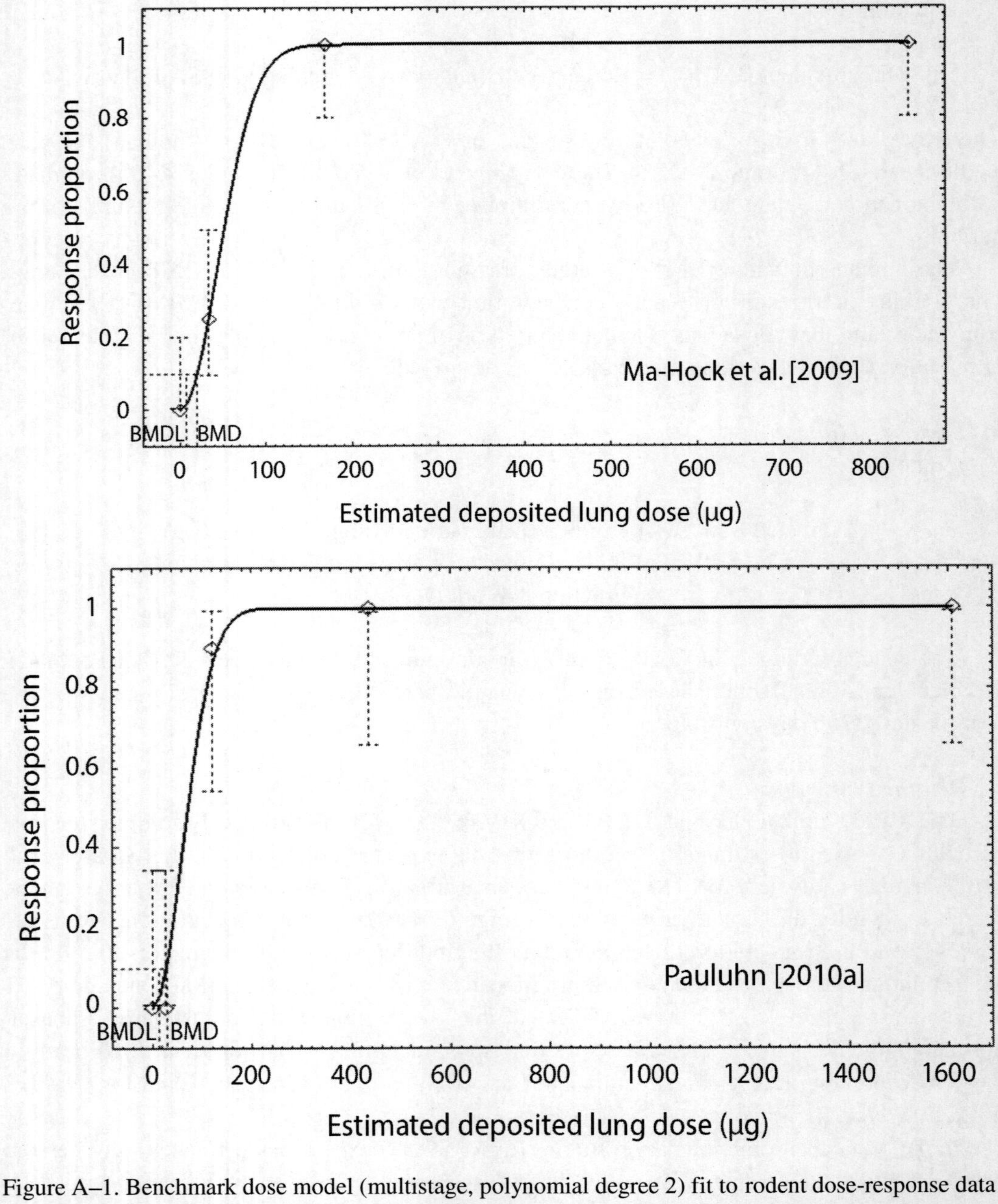

Figure A–1. Benchmark dose model (multistage, polynomial degree 2) fit to rodent dose-response data from the two subchronic inhalation studies of MWCNT in rats: Ma-Hock et al. [2009], response: granulomatous inflammation, Pauluhn [2010a], response: alveolar septal thickening, minimal or greater. *P* values are 0.99 for Ma-Hock et al. [2009] and 0.88 for Pauluhn [2010a].

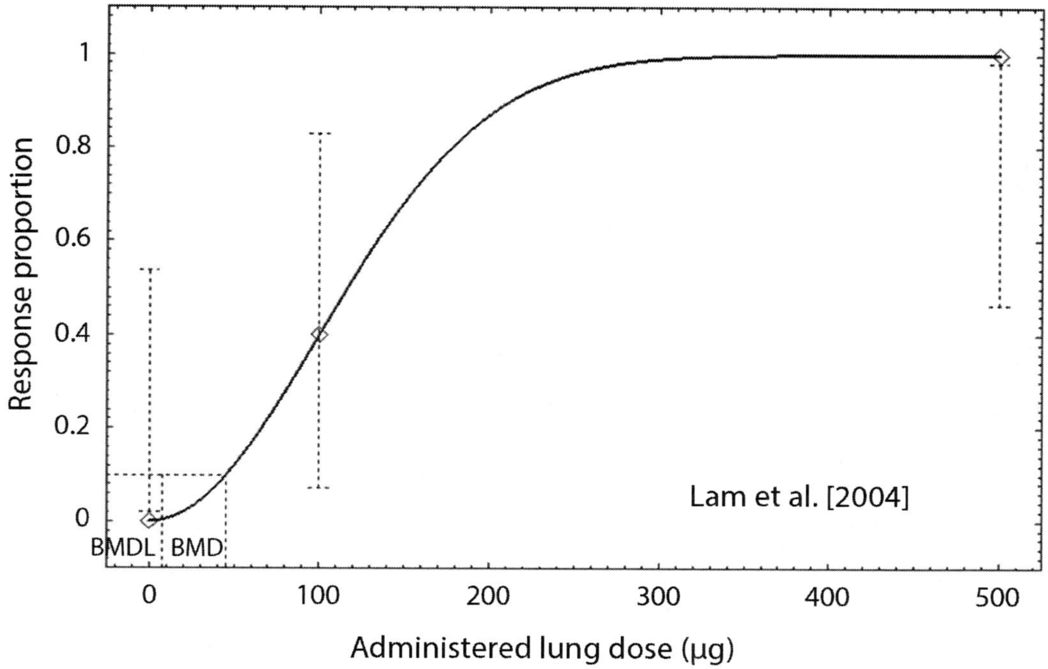

Figure A–2. Benchmark dose model (multistage, polynomial degree 2) fit to rodent dose-response data from shortterm studies with dichotomous response: Lam et al. [2004] (SWCNT, mouse, intratracheal instillation; response: lung granulomas) (*P* value is 0.35).

A.3. Results

A.3.1. Benchmark Dose and Working Lifetime Exposure Estimates

The estimates of the rodent BMD(L)s, the human-equivalent BMD(L)s, and the associated working lifetime 8-hr TWA exposure concentrations (MLE and 95% LCL)—called the benchmark concentration (BMC) and the BMCL (95% LCL of the BMC)—are shown in Tables A–3 through A–5. All dose-response models used in this risk assessment provided adequate fit ($P > 0.05$) to the rodent data for BMD(L) estimation (*P* values for the Pearson X^2 goodness of fit test shown in Tables A–3 through A–5).

In Table A–3, the BMD(L) and BMC(L)* estimates are based on the IT, PA, or shortterm inhalation exposure studies of SWCNT or MWCNT with continuous response measures. Lung responses in rodents were evaluated at 32 to 60 days after first exposure. Rodent dose is the administered (IT or PA) or estimated deposited dose (inhalation). The BMR is the specified adverse lung response at 1.1 standard deviations above the estimated rodent control mean response (i.e., alveolar connective tissue thickness or amount of hydroxyproline) (as explained in Section A.2.3.2). Considerably higher 8-hr TWA concentrations are estimated based on the endpoint of lung hydroyxproline amount [Muller et al. 2005] compared with those based on the alveolar connective tissue thickness endpoint, which is a more sensitive (earlier) indicator of fibrosis [Mercer et al. 2008].

* Abbreviation for both BMC and BMCL estimates.

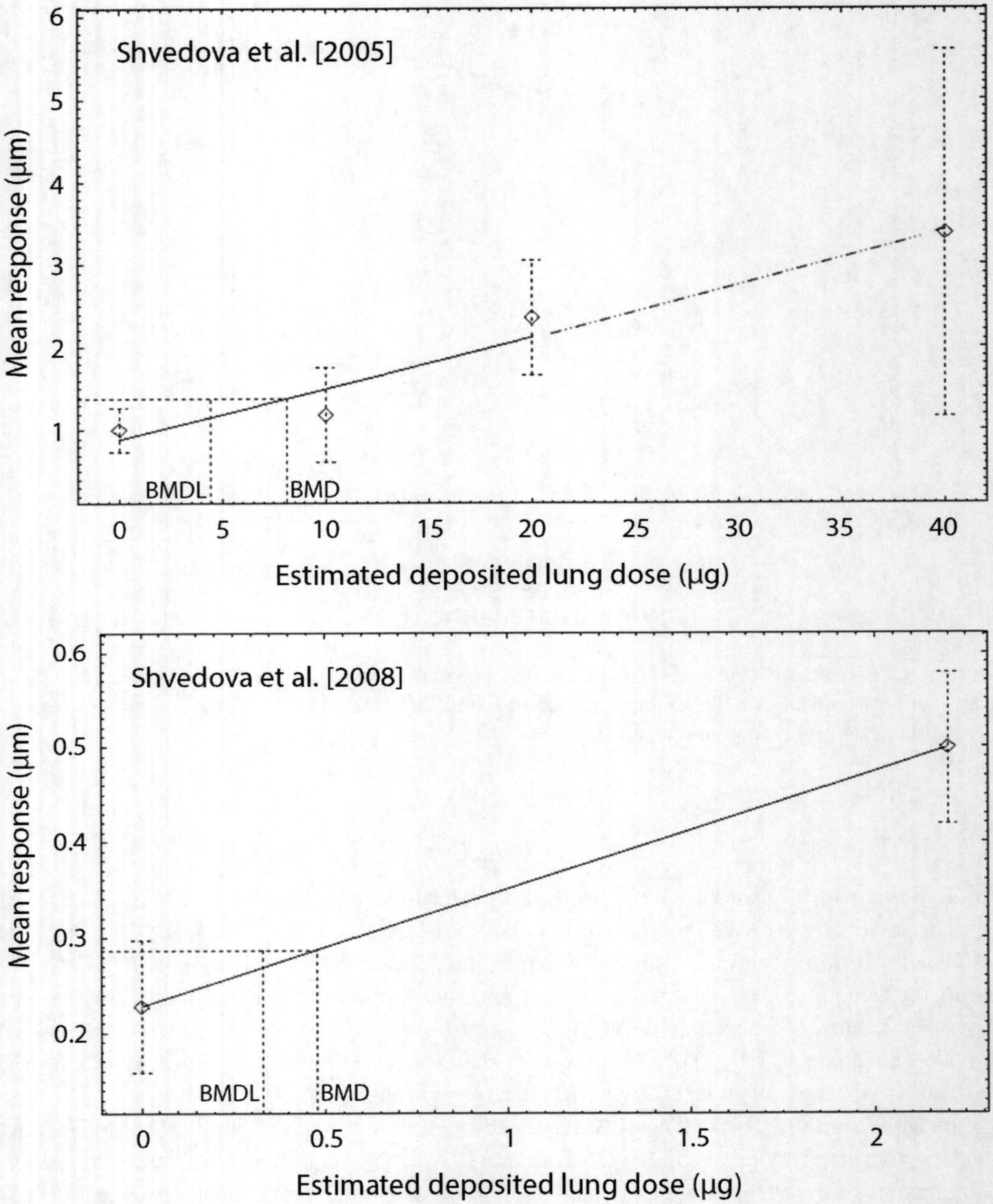

Figure A–3 (a). Benchmark dose model fit to rodent dose-response data from short-term studies with continuous response: Shvedova et al. [2005] (SWCNT, mouse, pharyngeal aspiration; response: alveolar connective tissue thickening) (*P* value is 0.23) (polynomial model, degree 2, coefficients restricted to non-negative—fit to all data except top dose group due to nonhomogeneous variance); and Shvedova et al. [2008] (SWCNT, mouse, inhalation; response: alveolar connective tissue thickening) (*P* value is not applicable, linear). Benchmark response level: 1.1 standard deviations about the control mean response.

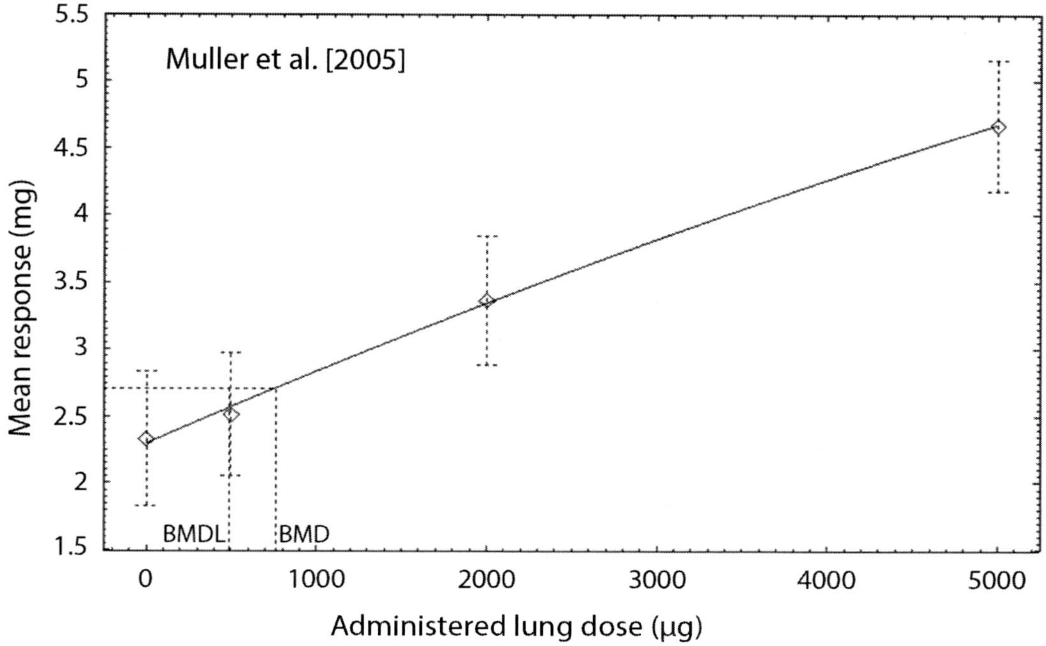

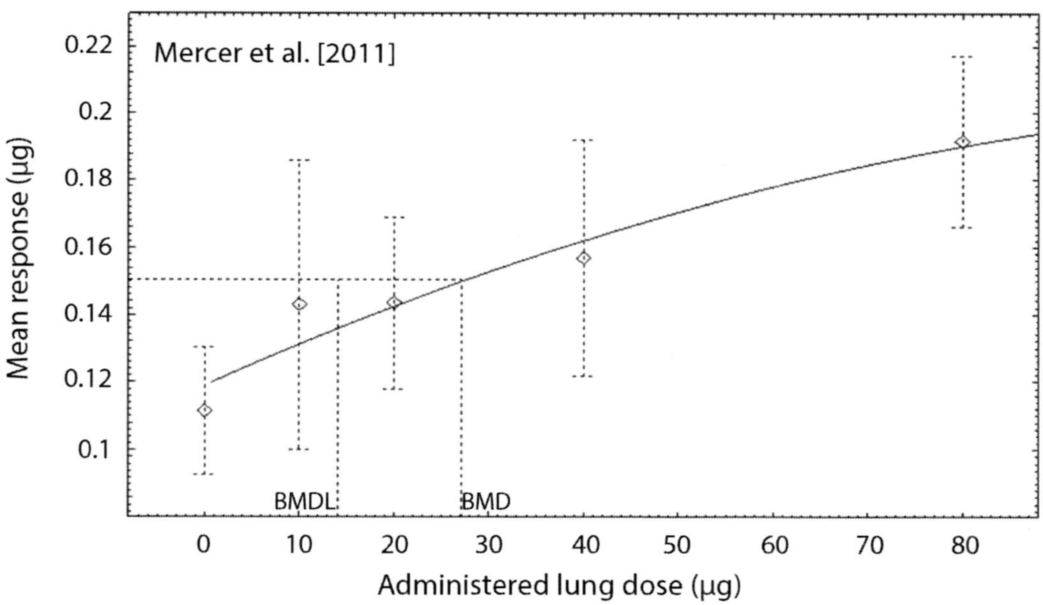

Figure A–3(b). Benchmark dose model fit to rodent dose-response data from shortterm studies with continuous response: Muller et al. [2005] (MWCNT, rat, intratracheal instillation; response: hydroxyproline amount) (*P* value is 0.67); Mercer et al. [2011] (MWCNT, mouse, pharyngeal aspiration response: alveolar connective tissue thickening) (*P* value is 0.35). Benchmark response level: 1.1 standard deviations about the control mean response.

Table A–3. Benchmark dose estimates* and associated human working lifetime airborne concentrations—continuous response data in rats or mice exposed to SWCNT or MWCNT by IT, PA,† or shortterm inhalation (dose metric: administered or estimated deposited lung dose)

Rodent study, CNT type, and response	Rodent		Human		Human working lifetime airborne concentration‡ (μg/m³)	
	BMD§, ¶ (μg/lung)	BMDL (μg/lung)	BMD (mg/lung)	BMDL (mg/lung)	BMC	BMCL
IT or PA						
Muller et al. [2005]—MWCNT (2% Al) Hydroxyproline amount (at 60 d) in rats	760	486	194	124	18	12
Shvedova et al. [2005]—SWCNT (0.2% Fe) Alveolar connective tissue thickness (at 60 d) in mice	7.8	6.5	14	12	1.8	1.5
Mercer et al. [2011]—MWCNT (0.3% Fe) Alveolar connective tissue thickness (at 60 d) in mice	27.1	14.1	50	27	4.7	2.5
Inhalation (5 hr/d, 4 d)						
Shvedova et al. [2008]—SWCNT (18% Fe) Alveolar connective tissue thickness (at 32 d) in mice	0.48	0.33	0.89	0.62	0.11	0.075

* Benchmark response level: 1.1 standard deviations above estimated control mean response [Crump 1995; US EPA [2010]; associated with a 10% increase in abnormal response (assumed greater than the 99th percentile of the distribution of control responses).

† IT—intratracheal instillation [Muller et al. 2005]; PA—pharyngeal aspiration [Shvedova et al. 2005] and [Mercer et al. 2011].

‡ 8-hr time weighted average (TWA) concentration associated with the human-equivalent BMD(L)s; BMC: maximum likelihood estimate of the benchmark concentration. BMCL: 95% lower confidence limit of the BMC.

§ BMD: estimated benchmark dose (maximum likelihood estimate). BMDL: estimated 95% lower confidence limit of the BMD; polynomial (degree 2) model [US EPA 2010].

¶ *P* values for the rodent dose response models: 0.67 for Muller et al. [2005], 0.23 for Shvedova et al. [2005], 0.41 for Mercer et al. [2011], and not applicable (linear) for Shvedova et al. [2008].

In Table A–4, the BMD(L) and BMC(L) estimates are based on the IT exposure study of SWCNT with dichotomous response measures. Lung responses were evaluated 90 days after the first exposure. The BMR is the 10% excess risk of the specified adverse lung response (proportion of rats with lung granulomas). Although Lam et al. [2005] report dose-response data for three different preparations of SWCNT (containing either 2% Fe, 27% Fe, or 26% Ni), the BMD(L) and BMC(L) estimates are provided only for the SWCNT with 2% Fe, which was the only dataset of the three reported by Lam et al. [2005] that was adequately fit by the BMD model (Table A–4).

Table A–4. Benchmark dose estimates* and associated human working lifetime airborne concentrations—dichotomous response data in rats or mice exposed to SWCNT by intratracheal instillation (dose metric: administered lung dose)

Rodent study, CNT type, and response	Rodent		Human		Human working lifetime airborne concentration† (μg/m³)	
	BMD‡, § (μg/lung)	BMDL (μg/lung)	BMD (mg/lung)	BMDL (mg/lung)	BMC	BMCL
Lam et al. [2004]— SWCNT (2% Fe) Granuloma (at 90 d) in mice	45	7.6	84	14	10	1.7

* Benchmark response level—10% excess (added) risk in exposed animal [US EPA 2010].

† 8-hr time weighted average (TWA) concentration associated with the human-equivalent BMD(L)s. BMC: maximum likelihood estimate of the benchmark concentration. BMCL: 95% lower confidence limit of BMC.

‡ BMD: estimated benchmark dose (maximum likelihood estimate). BMDL: estimated 95% lower confidence limit of the BMD; multistage (polynomial degree 2) model [US EPA 2010].

§ *P* value: 1.0

Table A–5 provides the BMD(L) and BMC(L) estimates based on the two subchronic inhalation studies of MWCNTs, which also report dichotomous response measures. Lung responses were evaluated at the end of the 13-week (91 d) exposure period. Rodent dose is either the total deposited lung dose or the retained lung dose at the end of exposure. The BMR is estimated as the 10% excess risk of the specified adverse lung response (granulomatous inflammation or alveolar septal thickening of histopathology grade 1 or higher). As expected, the estimates based on deposited lung dose are lower than those based on the retained lung dose, because the assumption of no clearance in the deposited lung dose results in a lower estimated 8-hr TWA concentration to attain the human-equivalent BMD(L) lung burdens. The estimates for MWCNT (with 9.6% Al^2O^3) based on the rat granulomatous in- flammation response are lower than those for MWCNT (Baytubes) (with 0.53% Co) based on the rat alveolar septal thickening response.

Table A–6 shows the animal and human BMD(L) estimates and equivalent working lifetime 8-hr TWA concentration estimates, BMC(L), associated with grade 2 (slight/mild) or higher lung responses in the subchronic inhalation studies, based on the estimated deposited lung dose. As expected, higher BMD(L)s and BMC(L)s are estimated from the histopathology grade 2 or higher lung responses (Table A–6) compared with those estimated from histopathology grade 1 (minimal) or higher responses (Table A–5) because more animals developed the grade 1 or higher response at a given dose (i.e., histopathology grade 1 or higher is a more sensitive response).

A.3.2. Comparison of Short- Term and Subchronic Dose-Response Data

Two studies of MWCNT (Baytubes) provided an opportunity to examine the effect of dose-rate on the same lung response measured at the same time point. Wistar rats were exposed by inhalation for either one 6-hr day [Ellinger-Ziegelbauer and Pauluhn 2009] or 13 weeks (6 hr/d, 5 d/wk) [Pauluhn 2010a]. Lung responses were examined in both studies at 13 weeks after the first exposure day. Histopathology severity scores for alveolar septal thickening were available for each study. The number of male rats with alveolar septal thickening (of minimal or higher grade) and the respective exposure concentrations are as follows:

- Ellinger-Ziegelbauer and Pauluhn [2009]: 1, 0, and 6 rats (6 total per group) at 0, 11.0, and 241.3 mg/m^3.
- Pauluhn [2010a, extended by personal communication]: 0, 0, 9, 10, 10 rats (10 total per group) at 0, 0.1, 0.45, 1.62, and 5.98 mg/m^3.

The dose metric used for this comparison was the deposited lung dose, estimated from MPPD 2.0 [CIIT and RIVM 2006] based on the particle size data (MMAD and GSD) and the rat exposure conditions reported in each study.

Table A–5. Benchmark dose estimates and associated working lifetime airborne concentrations—grade 1 or higher severity of lung responses in rats after subchronic inhalation of MWCNT (dose metric: estimated deposited or retained dose in lungs)

Rodent study and response*	Rodent		Human		Human working lifetime airborne concentration‡ (µg/m^3)	
	BMD† (µg/lung)	BMDL (µg/lung)	BMD (mg/lung)	BMDL (mg/lung)	BMC	BMCL
Deposited lung dose (assuming no clearance)						
Ma-Hock et al. [2009] Granulomatous inflammation	21	8.1	5.4	2.1	0.51	0.19
Pauluhn [2010a] Alveolar septal thickening	28	14	7.2	3.5	0.77	0.38
Retained lung dose (assuming normal clearance)						
Ma-Hock et al. [2009] Granulomatous inflammation	11	3.8	2.7	0.97	2.7	1.0
Pauluhn [2010a] Alveolar septal thickening	14	6.5	3.6	1.7	4.2	1.9

*Histopathology grade 1 (minimal) or higher severity. Benchmark response level—10% excess (added) risk in exposed animals.

† BMD: estimated benchmark dose (maximum likelihood estimate). BMDL: estimated 95% lower confidence limit of the BMD; multistage (polynomial degree 2) [US EPA 2010]. *P* values for the rodent dose-response models: 0.99 for Ma-Hock et al. [2009] and 0.88 for Pauluhn et al. [2010a] (deposited dose); 1.0 for Ma-Hock et al. [2009] and 0.94 for Pauluhn [2010a] (retained dose).

‡ 8-hr time weighted average (TWA) concentration associated with the human-equivalent BMD(L)s. BMC: maximum likelihood estimate of the benchmark concentration; BMCL: 95% lower confidence limit of the BMC.

Table A–6. Benchmark dose estimates and associated working lifetime airborne concentrations—grade 2 or higher severity of lung responses in rats after subchronic inhalation of MWCNT (dose metric: estimated deposited or retained dose in lungs)

Rodent study and response*	Rodent		Human		Human working lifetime airborne concentration‡ ($\mu g/m^3$)	
	BMD† (μg/lung)	BMDL (μg/lung)	BMD (mg/lung)	BMDL (mg/lung)	BMC	BMCL
Deposited lung dose						
Ma-Hock et al. [2009] Granulomatous inflammation	44	29	11	7.4	1.0	0.69
Pauluhn [2010a] Alveolar septal thickening	235	120	60	31	6.4	3.3
Retained lung dose						
Ma-Hock et al. [2009] Granulomatous inflammation	24	16	6.3	4.0	6.2	4.0
Pauluhn [2010a] Alveolar septal thickening	150	66	38	17	44	19

* Histopathology grade 2 is slight [Ma-Hock et al. 2009] or slight/mild [Pauluhn 2010a] severity (proportion of animals with that response). Benchmark response level—10% excess (added) risk in exposed animals.

† BMD: estimated benchmark dose (maximum likelihood estimate). BMDL: estimated 95% lower confidence limit of the BMD; multistage (polynomial degree 2) [US EPA 2010]. P values for the rodent dose-response models: 0.67 for Ma-Hock et al. [2009] and 0.98 for Pauluhn et al. [2010a] (deposited dose); 0.76 for Ma-Hock et al. [2009] and 0.99 for Pauluhn [2010a] (retained dose).

‡ 8-hr time weighted average (TWA) concentration associated with the human-equivalent BMD(L)s. BMC: maximum likelihood estimate of the benchmark concentration; BMCL: 95% lower confidence limit of the BMC.

To evaluate whether these data [Pauluhn 2010a; Ellinger-Ziegelbauer and Pauluhn 2009] can be described by the same dose-response relationship, a multistage (polynomial degree 2) model [US EPA 2010] was fit to the combined data (Figure A–4). This model provided adequate fit to the data ($P = 0.37$), suggesting that these data can be described by the same dose-response model using the estimated total deposited lung dose, regardless of the dose rate differences (i.e., dose administered in 1 day vs. during 91 days). This finding is consistent with the impaired clearance and biopersistence of the deposited MWCNT in the rat lungs at these doses as shown in Pauluhn [2010a].

A.3.3. Estimated Excess Risks of Working Lifetime Exposures to Low-Mass Concentrations of MWCNT

Standard risk assessment approaches for non-cancer endpoints have typically assumed a threshold model, with extrapolation beyond the point of departure based on uncertainty factors, e.g., US EPA [1994]. NRC [2009] and others have recommended using risk-based low-dose extrapolation for non-cancer endpoints. NIOSH practice has also included risk-based low-dose extrapolation for non- cancer endpoints. In the absence of information on the shape of the dose-response relationship in the low-dose region, assumptions can include linear and nonlinear model-based extrapolation. Linear extrapolation is the most protective (i.e., unlikely to underestimate the risk). However, the actual risk could be much lower, including zero.

Table A–7. Working lifetime percent, excess risk estimates of low-mass concentrations of CNT associated with minimal (grade 1) lung effects

Subchronic inhalation study in rats	Working lifetime 8-hr TWA airborne concentration (μg/m³)	Working lifetime excess risk (%)*	
		Maximum likelihood estimate (MLE)	95% Upper confidence limit of MLE
Deposited lung burden (assumes no clearance)			
Ma-Hock et al. [2009]	1	33	54
	2	80	96
	7	>99	>99
Pauluhn [2010]	1	16	30
	2	50	72
	7	>99	>99
Retained lung burden (assumes normal clearance)			
Ma-Hock et al. [2009	1	3.7	10
	2	7.4	20
	7	49	73
Pauluhn [2010]	1	2.4	5.3
	2	4.8	10
	7	25	42

* 45-year working lifetime; estimated from multistage model (degree 2) [US EPA 2010] for exposures greater than 10% BMC(L) and by linear extrapolation from the 10% BMC(L) in Table A–5 for lower exposures.

Table A–8. Working lifetime percent, excess risk estimates of low-mass concentrations of CNT associated with slight/mild (grade 2) lung effects

Subchronic inhalation study in rats	Working lifetime 8-hr TWA airborne concentration (μg/m³)	Working lifetime excess risk (%)*	
		Maximum likelihood estimate (MLE)	95% Upper confidence limit of MLE
Deposited lung burden (assumes no clearance)			
Ma-Hock et al. [2009]	1	10	16
	2	31	44
	7	99	>99
Pauluhn [2010]	1	1.6	3.0
	2	3.1	6.1
	7	12	24
Retained lung burden (assumes normal clearance)			
Ma-Hock et al. [2009	1	1.6	2.5
	2	3.2	5.0
	7	12	21
Pauluhn [2010]	1	0.23	0.53
	2	4.5	1.0
	7	1.6	3.7

* 45-year working lifetime; estimated from multistage model (degree 2) [US EPA 2010] for exposures greater than 10% BMC(L) and by linear extrapolation from the 10% BMC(L) in Table A–6 for lower exposures.

Low-dose linear extrapolation of the working life-time-equivalent 10% excess risk estimates in Table A–5 (deposited dose assumption) results in BMC (BMCL) estimates of 0.051 (0.019) μg/m^3 or 0.077 (0.038) μg/m^3 associated with 1% excess risk (Ma-Hock et al. [2009] or Pauluhn [2010a], respectively). The corresponding BMC (BMCL) estimates associated with 0.1% excess risk are 0.0051 (0.0019) μg/m^3 or 0.0077 (0.0038) μg/m^3. Multistage model-based estimates are higher for the BMCs, but nearly identical for the BMCLs: 0.16 (0.019) μg/m^3 or 0.24 (0.042) μg/m^3 associated with 1% excess risk; and 0.050 (0.0020) μg/m^3 or 0.075 (0.0042) μg/m^3 associated with 0.1% excess risk (Ma-Hock et al. [2009] or Pauluhn [2010a], respectively).

Tables A–7 and A–8 provide working lifetime excess risk estimates of early stage-lung effects (minimal or higher histopathology grade of granulomatous inflammation or alveolar septal thickening) associated with 1, 2, or 7 μg/m^3 as an 8-hr TWA concentration. These concentrations were selected as possible limits of quantification (LOQs) that were under evaluation for the analytical method to measure airborne CNT in the workplace (NIOSH method 5040). These estimates are based on lung dose estimates assuming either total deposited lung dose (no clearance) or retained dose (normal, spherical particle-based clearance). Risk estimates are higher for the no clearance assumption than those assuming normal clearance, within either the minimal (grade 1) (Table A–7) or slight/mild (grade 2) (Table A–8) lung responses. These excess (exposure-attributable) risk estimates were derived from the multistage (degree 2) model fit to the rat subchronic dose-response data, or by linear extrapolation below the 10% BMC(L) estimates shown in Tables A–5 and A–6.

A.4. Discussion

NIOSH conducted a quantitative risk assessment of CNTs by evaluating dose-response data of earlystage adverse lung effects in rats and mice exposed to several types of SWCNT or MWCNT (with different metal contaminants), by several routes of exposure (inhalation, PA, or IT), and duration of exposure (single day or subchronic) and post-exposure period (up to 26 weeks). Because of the different study designs and response endpoints used in the rodent studies, limited information was available to evaluate the extent to which the differences in the risk estimates across studies are due to differences in the CNT material or are attributable to other study differences. Some evidence indicates that CNT with certain metals (nickel, 26%) [Lam et al. 2004] or with higher metal content (18% vs. 0.2% Fe) [Shvedova et al. 2008] are more toxic and fibrogenic. However, some studies have shown that both unpurified and purified (low metal content) CNT were associated with early-onset and persistent pulmonary fibrosis at relatively low-mass doses [Shvedova et al. 2005, 2008]. The LOAELs for MWCNT (containing either 9.6% Al_2O_2 or 0.5% Co) were 0.1 mg/m^3 [Ma-Hock et al. 2009] and 0.4 mg/ m^3 [Pauluhn 2010a], which are more than an order of magnitude lower than the LOAEL of 7 mg/m^3 for ultrafine carbon black [Elder et al. 2005] in the same animal species and study design (13-week inhalation studies in rats, although with different strains, Wistar (male and female) [Pauluhn 2010a] and F-344 (female) [Elder et al. 2005]).

Because no chronic animal studies or epidemiological studies of workers producing or using CNT have been published to date, the best available data for risk assessment were the subchronic inhalation studies of MWCNT in rats [Ma-Hock et al. 2009; Pauluhn 2010a]. For SWCNT, no subchronic studies were available, and several shortterm studies (IT, PA, or inhalation exposure) in rats or mice provide the only available dose-response data for either SWCNT [Lam et al. 2004; Shvedova et al. 2005, 2008] or for other types of MWCNT (with different metal content) [Muller et al. 2005; Mercer et al. 2011] (Table A–1).

All of these studies reported inflammatory, granulomatous, and/or fibrotic lung effects of relevance to human health risk assessment. These lung effects in the animal studies were relatively earlystage and were not reversible after exposure ended (up to approximately 6 months post-exposure [Pauluhn 2010a]). In the studies with multiple post-exposure followup times, the amount of pulmonary fibrosis persisted or progressed with longer followup [Shvedova et al. 2005, 2008; Mercer et al. 2008; Porter et al. 2010; Pauluhn 2010a]. One of the measures of pulmonary fibrosis used in the shortterm studies [Shvedova et al. 2005, 2008; Mercer et al. 2008, 2011]—alveolar epithelial cell thickness (due to increased collagen deposition associated with CNT mass lung dose)—was also used to develop the EPA ozone standard. This response endpoint was selected by EPA as the adverse lung response for cross-species dose-response extrapolation, because it indicates "fundamental structural remodeling" [US EPA 1996; Stockstill et al. 1995].

The excess risk estimates based on the subchronic and shortterm studies of MWCNT and SWCNT suggest that workers are at >10% excess risk of developing earlystage adverse lung effects (pulmonary inflammation, granulomas, alveolar septal thickening, and/or fibrosis) if exposed for a working lifetime at the estimated upper LOQ of 7 μg/m^3 based on NIOSH Method 5040 for measuring the airborne concentration of CNT [NIOSH 2010a] (Appendix C; Tables A–3 through A–8). Working lifetime airborne concentration (8-hr TWA) estimates of 0.51–4.2 μg/m^3 MLE and 0.19–1.9 μg/m^3 95% LCL were associated with a 10% excess risk of earlystage lung effects (histopathology grade 1 minimal or higher) based on the subchronic inhalation studies (Table A–5). For histopathology grade 2 (slight [Ma-Hock et al. 2009] or slight/mild [Pauluhn 2010a]), the working lifetime 8-hr TWA concentrations associated with an estimated 10% excess risk are 1.0 to 44 μg/m^3 MLE and 0.69 to 19 μg/m^3 95% LCL (Table A–6).

As discussed in Section A.2.3, the 10% BMDL estimates are a typical POD for extrapolation to lower risk. NIOSH does not consider 10% or greater excess risk levels of these earlystage lung effects to be acceptable if equivalent effects were to occur in workers as a result of working lifetime exposures to CNT. Linear extrapolation by application of uncertainty factors (e.g., Table A–14) would result in lower 8-hr TWA concentrations. However, the lowest LOQ of NIOSH Method 5040 (1 μg/m^3) is the best that can be achieved at this time in most workplaces and is similar to or greater than the 8-hr TWA concentrations estimated to be associated with 10% excess risk of minimal (grade 1) effects (Table A–7). Some of the risk estimates are less than 10% at the LOQ of 1 μg/m^3 (8-hr TWA), in particular those based on the slight/mild (grade 2) rat lung effects and assumed normal clearance (Table A–8).

Although uncertainties and limitations exist in these animal studies, the evidence supports the health-based need to reduce exposures below 1 μg/ m^3. These risk estimates indicate the need for research to develop more sensitive measurement methods for airborne CNT in the workplace, to demonstrate effective exposure control, and to evaluate the need for additional risk management measures such as the use of respirators and other personal protective equipment and medical screening (Section 6, Appendix B). Chronic bioassay data are also needed to reduce the uncertainty concerning the potential for chronic adverse health effects from longterm exposure to CNT. Evaluation of the factors that influence the risk estimates and the areas of uncertainty are discussed below.

A.4.1. The Use of Shortterm Data to Predict Longer-Term Response

Several factors suggest that in the absence of chronic data these shortterm and subchronic animal data may be reasonable for obtaining initial estimates of the risk of human noncancer lung effects from exposure to CNT. First, some fraction of CNT that deposit in the lungs are likely to be biopersistent based on studies in animals [Muller et al. 2005; Deng et al. 2007; Elgrabli et al. 2008b; Mercer et al. 2009; Pauluhn 2010a, b] and studies of other poorly soluble particles in human lungs [ICRP 1994; Kuempel et al. 2001; Gregoratto et al. 2010]. Second, the pulmonary fibrosis developed earlier and was of equal or greater severity than that observed from exposure to the same mass dose of other inhaled particles or fibers (silica, carbon black, asbestos) examined in the same study [Shvedova et al. 2005; Muller et al. 2005]. Third, the adverse lung responses persisted or progressed after the end of exposure up to 90 days after a single- or multiple-day exposure to SWCNT or MWCNT [Lam et al. 2004; Muller et al. 2005; Shvedova et al. 2005, 2008; Ellinger-Ziegelbauer and Pauluhn 2009; Porter et al. 2010] or 26 weeks after a 13-week inhalation exposure to MWCNTs (Baytubes) [Pauluhn 2010a].

There is uncertainty in estimating working-lifetime health risk from either subchronic or shortterm animal studies, and perhaps from the shorter-term studies. The strength of the subchronic inhalation studies is that they provide exposure conditions that are more similar to those that may be encountered by workers exposed to airborne CNT. However, there is some uncertainty about the deposited and retained dose in the rat lungs (see Section A.6.1 for sensitivity analysis of the lung dose estimates). In the PA or IT studies, the administered lung dose is known, although the pattern of lung deposition (especially for IT administration) may differ from that of inhalation. The subchronic inhalation studies and some of the PA studies include multiple doses, which can provide better information about the shape of the dose-response relationship. However, in the subchronic studies, steep dose-response relationships were observed for lung response proportions based on histopathology score, reaching 100% response for minimal or higher severity (grade 1) (Figure A–1). Although the data are sparse in the low dose region (near a 10% response level), the BMD(L) estimates are generally similar to the LOAEL and NOAEL values reported in those studies (Section A.6.2 and Table A–12).

A comparison of data from 1-day and 13-week inhalation exposures in rats [Ellinger-Ziegelbauer and Pauluhn 2009; Pauluhn 2010a], indicates that the dose-response relationship was consistent despite the differences in dose-rate in those two studies (Figure A–4). This finding indicates that it may be reasonable to assume that the dose-response relationships for the IT, PA, and shortterm inhalation exposure studies would be consistent with the subchronic study results if the same response is examined at the same time point, although additional study is needed to confirm this finding. The BMC(L) estimates among the subchronic and shortterm studies (Tables A–3 through A–5) are reasonably consistent.

A.4.2. Physical-Chemical Properties and Metal Content

There are limited data to evaluate the role of physical-chemical properties of CNT on the lung responses. Although the dose estimates vary for the earlystage lung effects in rats and mice (and in the human-equivalent concentrations (Tables A–3 through A–6), all estimates are relatively low mass concentrations. It is difficult to tease out the CNT-specific factors affecting these estimates from those due to the other study differences (e.g., exposure route, duration, animal species, lung response measures).

The two subchronic inhalation studies of MWCNT [Ma-Hock et al. 2009; Pauluhn 2010a], based on the same study design (13 week inhalation) and animal species/strain (Wistar rats), facilitates comparison. Different types of MWCNT and different generation methods for aerosolizing exposures were used in each study, although the primary particle sizes reported were similar—approximately 10 nm in width and 0.1–10 μm in length, with specific surface area of approximately 250–300 m^2/g [Ma-Hock et al. 2009; Pauluhn 2010a]. The aerodynamic diameter (and resulting alveolar deposition fraction) estimates were also fairly similar (Table A–2); yet the bulk densities differed (approximately 0.04 and 0.2 g/ml, respectively, in Ma-Hock et al. [2009] and Pauluhn [2010a]). The metal content also differed, with 9.6% Al2O3 in the MWCNT in the Ma-Hock et al. [2009] study vs. 0.5% Co in the MWCNT (Baytubes) in the Pauluhn [2010a] study. The lung responses differed both qualitatively and quantitatively, including "pronounced granulomatous inflammation, diffuse histiocytic and neutrophilic inflammation, and intraalveolar lipoproteinosis" with a LOAEL of 0.1 mg/m^3 in Ma-Hock et al. [2009], vs. "inflammatory changes in the bronchioloalveolar region and increased interstitial collagen staining" with a LOAEL of 0.45 mg/m^3 [Pauluhn 2010a]. Yet, both MWCNT studies reported LOAELs that are lower by more than an order of magnitude compared to the LOAEL (7 mg/m^3) reported in a 13-week inhalation study of ultrafine carbon black [Elder et al. 2005].

A recent study provides a quantitative comparison of the effects of SWCNT and MWCNT on pulmonary interstitial fibrosis [Mercer et al. 2011]. In this study, MWCNTs were administered to mice by pharyngeal aspiration at several different doses (0 [control], 10, 20, 40, or 80 μg); the lung tissues (stained for collagen using Sirius red) were examined at 56 days post-exposure. At the 80-μg dose of MWCNT, the average thickness of the alveolar interstitial connective tissue was significantly increased at 28 days, and a progressive increase in thickness was observed at 56 days. The 40-μg MW- CNT dose group also showed a significant increase in the interstitial connective tissue thickness at 56 days. These data were compared with those of an earlier study of SWCNT [Mercer et al. 2008] using the same study design. The individual MWCNTs had a mean diameter of 49 nm and a mean length of 3.9 μm. The individual SWCNTs were 1–4 nm in diameter and several hundred nanometers in length. Both SWCNT and MWCNT were rapidly incorporated into the alveolar interstitial spaces (within 1 hour individual CNT or small clumps of CNT were observed), although the percentage of the administered SWCNT observed in the alveolar interstitium (~90%) was much higher than that for MWCNT (~8%). After accounting for the differences in the target tissue dose, SWCNTs were still ~8.5–fold more fibrogenic than MWCNTs. However, the surface area of SWCNT was ~20-fold greater per unit mass than that of MWCNTs (508 m^2/g for SWCNT vs. 26 m^2/g for MWCNT), suggesting that the greater fibrogenic potency of SWCNT may be due to its greater surface area. When the lung response was evaluated per unit CNT surface area dose, SWCNT was no longer more potent, and the MWCNT were 2.5-fold more potent on a surface area basis. There is uncertainty about the degree of dispersion (and hence available surface) of these materials *in vivo*, which precludes assigning exact potency factors [Mercer et al. 2011]. However, these findings suggest that the greater fibrotic potency of SWCNT on a mass basis is likely due to its greater surface area available to react with lung tissue.

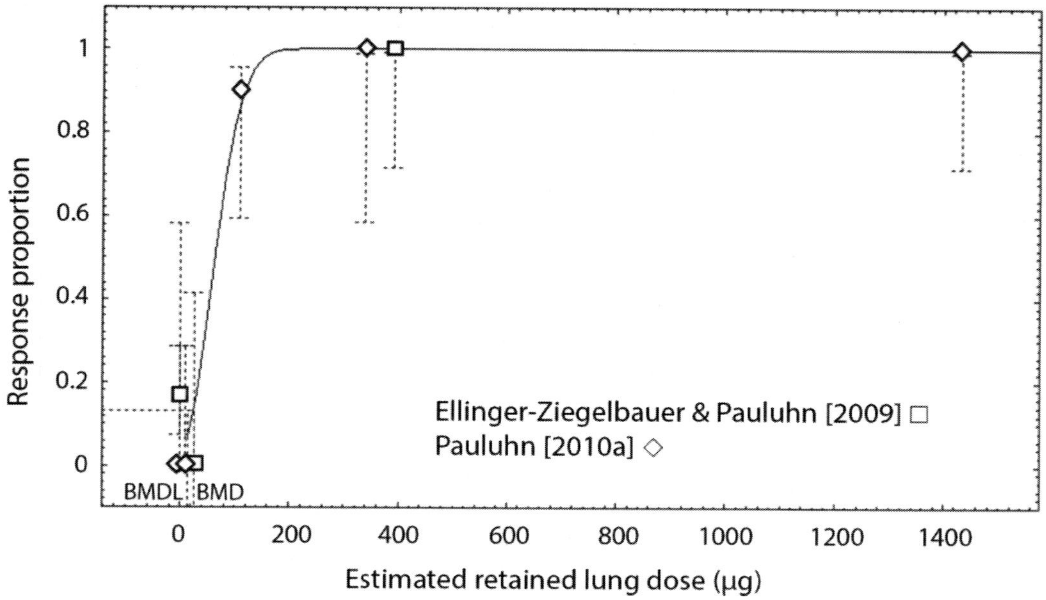

Figure A–4. Dose-response relationship between estimated retained lung doses of MWCNT (Baytubes) based on cobalt-tracer measurements and earlystage pulmonary fibrosis (proportion of rats with minimal or greater alveolar interstitial thickening) examined at 13 weeks, following either a 1-day [Ellinger-Ziegel- bauer and Pauluhn 2009] or 13-week inhalation exposure [Pauluhn 2010a]. Dose groups include n=10 [Pauluhn 2010a] or n = 6 [Ellinger-Ziegelbauer and Pauluhn 2009]. Data were fit with a multistage (polynomial degree 2) model in BMDS 2.2 [US EPA 2010]. Error bars are the 95% confidence limits.

Comparison of other CNT types and metal content is generally impeded by differences in study design. In one of the few studies to investigate CNT with different metal content, Lam et al. [2004] reported lung granuloma and inflammation responses in mice administered IT doses of SWCNT containing either 2% Fe, 27% Fe, or 26% Ni. The number of mice developing granulomas by group (each containing 5 mice) were the following:

- 0.1 mg dose: 2 (2% Fe); 5 (27% Fe); and 0 (26% Ni)
- 0.5 mg dose: 5 (2% Fe); 5 (27% Fe); and 5 (26% Ni)

In addition, three mice died in the first week in the 0.5 mg dose of the 26% Ni group.

Because of the sparse data and the steep dose-response relationship, only the SWCNT containing 2% Fe were adequately fit by the BMDS model. The high mortality rate in mice exposed to the SWCNT containing Ni suggests this material is highly toxic. The greater response proportion in the mice exposed to 0.1 mg SWCNT with 27% Fe (5/5) compared with rats exposed to the same dose of SWCNT with 2% Fe (2/5) suggests that the CNT with higher Fe content are more toxic than CNT with lower Fe content.

In Shvedova et al. [2005, 2008], higher iron content was also associated with greater lung response and thus lower BMD(L) estimates. The BMD(L) estimates for SWCNT with 18% Fe were lower than those for SWCNT with 0.2% Fe (Table A–3), even though the post-exposure time was longer (60 vs. 28 days) for the 0.2% Fe SWCNT [Shvedova et al. 2005,

2008]. All types of CNT (including SWCNT and MWCNT, purified or unpurified, and with various types and percentages of metals) were of similar or greater potency (i.e., similar or greater lung responses at the same mass dose) in these animal studies compared to the other types of particles or fibers tested (asbestos, silica, ultrafine carbon black) [Lam et al. 2004; Muller et al. 2005; Shvedova et al. 2005, 2008].

A.4.3. Lung Dose Estimation

In any CNT risk assessment, there may be greater uncertainty in the estimated lung dose of respirable CNT than there is for spherical airborne particles for which lung dosimetry models have been developed and validated. Evaluations have not been made on the influence of particle characteristics (e.g., shape and density) on the inhalability and deposition of CNT in the human respiratory tract, and on the clearance or biopersistence of CNT. However, the available data on the aerodynamic size of CNT provides an initial estimate (based on validated models for spherical particles) of the deposited mass fraction of airborne CNT in the human respiratory tract, and specifically in the alveolar (gas exchange) region. The clearance rate of CNT from the lungs may be more uncertain than the deposition efficiency, as animal studies indicate that CNT clearance becomes impaired in rat lungs at lower mass doses than for larger particles of greater density [Pauluhn 2010a, b]. The NIOSH risk assessment helps to characterize this uncertainty by providing bounds on the range of possible lung dose estimates, from assuming normal clearance to assuming no clearance of the deposited CNT. This approach also provides a framework for introducing improved dose estimates when validated lung dosimetry models for CNT become available.

The assumptions used in the lung dose estimation have a large influence on the animal and human-equivalent BMD(L) or BMC(L) estimates (Tables A–5 and A–6), as well as on the estimated human-equivalent NOAEL (Section A.6.3). The rat BMD(L) estimates based on the estimated retained lung dose after subchronic inhalation exposure in rats are lower than those based on the estimated deposited lung dose (Table A–5). This is because the retained dose estimates allow for some lung clearance to occur during the 13-week exposure in rats, and a lower dose estimate is therefore associated with a given fixed response proportion. The human-equivalent BMD(L) estimates based on retained dose are also lower because they are proportional to the rat BMD(L)s (i.e., calculated based on the ratio of the human to rat alveolar epithelial cell surface area). However, the working lifetime 8-hr TWA concentrations, BMC(L)s, based on the estimated retained lung doses are higher than those based on the estimated deposited lung dose. This is because the retained dose estimates (which assume some particle clearance from workers' lungs during the 45 years of exposure), require a higher inhaled airborne concentration to reach the estimated human-equivalent BMD(L) lung doses.

The estimated deposited lung dose of CNT (assuming no clearance) may overestimate the actual CNT lung dose, given that the shortterm kinetic data have shown some CNT clearance in rats and mice [Muller et al. 2005; Deng et al. 2007; Elgrabli et al. 2008b; Mercer et al. 2009; Pauluhn 2010a, b]. On the other hand, the estimated retained lung dose of CNT, based on models for poorly soluble spherical particles, may underestimate the retained CNT lung burden, given that overloading of rat lung clearance has been observed at lower mass doses of MWCNT (Baytubes) than for other poorly soluble particles [Pauluhn 2010a,b]. Thus, although there is uncertainty in the deposition and retention of CNT in the animal and human lungs, the deposited and retained lung dose estimates reported in this risk assessment may represent reasonable upper and lower bounds of the actual lung doses.

A.4.4. Critical Effect Level Estimates

The response endpoints in these animal studies of CNT are all relatively earlystage effects. Although these effects were persistent or progressive after the end of exposure in some studies, there was no information on whether these responses were associated with adverse functional effects. More advanced-stage responses (grade 2 or higher severity on histopathology examination) were also evaluated, and as expected, these responses resulted in lower risk estimates (Table A–6). It is expected that exposure limits derived from these early response data would be more protective than those based on frank adverse effects. On the other hand, because of the lack of chronic studies, there is considerable uncertainty about the potential chronic adverse health endpoints.

The excess risk estimates at the lower LOQ (1 μg/ m^3) are considerably lower than those at the upper LOQ (7 μg/m^3) of NIOSH Method 5040, for either minimal (TableA–7) or slight/mild (Table A–8) lung effects based on the rat subchronic inhalation data. The range in the estimates in Table A–7 and A–8 reflects the low precision in the animal data and the uncertainty about CNT retention in the lungs. There is also uncertainty about the relationship between the lung dose and response, including whether there is a threshold. For example, for slight/mild lung effects (Table A–8), the actual risk could be as low as zero or as high as 16% at the REL of 1 μg/m^3.

NIOSH utilized BMD modeling methods to estimate the critical effect level (i.e., the dose associated with the critical effect or benchmark response) in order to provide a standardized method for risk estimation across studies. In contrast, the NOAEL-based approaches do not estimate risk, but may assume safe exposure or zero risk below the derived OEL. BMD modeling also uses all of the dose-response data, rather than only a single dose for a NOAEL or LOAEL, and takes appropriate statistical account of sample size, unlike NOAEL-based approaches. However, the BMD modeling options for some of these CNT data were limited because of sparse data, and the dose groups with 100% response (observed in the subchronic inhalation studies) contribute little information to the BMD estimation. A common challenge in risk assessment is defining a biologically relevant response for continuous endpoints, which was also encountered in this risk assessment. A standard practice of using a statistical definition of the benchmark response was used for the continuous BMD estimation in the absence of data on the functional significance of the earlystage pulmonary inflammation and fibrotic responses (Section A.2.3.2).

For CNT, as with other chemicals, there is uncertainty in whether a NOAEL or a BMDL from a shortterm or subchronic study in animals would also be observed in a chronic study. For example, in the Pauluhn [2010a] study, 0.1 mg/m^3 was the NOAEL based on subchronic inhalation exposure in rats, but there was some indication that lung clearance overloading may have already begun (i.e., retention half-time about two-fold higher than normal, although imprecision in the low dose measurement was noted) [Pauluhn 2010a, b]. A comparison of the BMD and the NOAEL estimates shows that these estimates are statistically consistent (Section A.6.2). Thus, there is uncertainty as to whether chronic exposure at 0.1 mg/m^3 might result in adverse lung effects that were not observed during subchronic exposure. It is also uncertain whether these subchronic effects (without additional exposure) would resolve with longer post-exposure duration (beyond the 26-week post-exposure period in the Pauluhn

[2010a] study). Yet, workers may be exposed to CNT daily for many years, e.g., up to a working lifetime. The NIOSH REL is intended to reduce the risk of lung disease from exposures to CNT and CNF up to a 45-year working lifetime.

A.4.5. Animal Dose-Response Data Quality

In the absence of epidemiological data for CNT, the two subchronic inhalation studies of two types of MWCNT, in addition to the shortterm studies of SWCNT and MWCNT, provide the best available dose-response data to develop initial estimates of the risk of earlystage adverse lung responses associated with exposure to CNT. The availability of animal dose-response data for different types of CNT—and the consistent low mass concentration BMC(L) estimates—suggests these risk estimates are relatively robust across a range of CNT types, including SWCNT or MWCNT, either purified or unpurified (containing different types and amounts of metal), dispersed or agglomerated. Although a formal comparison of the potency of the different CNT is not feasible because of differences in study design, these studies consistently show that relatively low-mass doses of CNT are associated with earlystage adverse lung effects in rats and mice. Consequently, the human-equivalent benchmark dose and working lifetime exposure estimates derived from these studies are also relatively low on a mass basis. The excess risk estimates of earlystage adverse lung responses to CNT generally indicate > 10% excess risk (lower 95% confidence limit estimates) at the upper LOQ (7 $\mu g/m^3$) of the measure ment method (NIOSH Method 5040) regardless of the CNT type or purification (Tables A–3 through A–5). Lower risks are estimated at the optimal LOQ (1 $\mu g/m^3$), depending on lung dose assumptions (Tables A–7 through A–8).

A more indepth analysis of specific areas of uncertainty in this CNT risk assessment is provided in Section A.6. This includes quantitative evaluation of the methods and assumptions used in the CNT risk assessment for the derivation of a REL.

A.5. Conclusion

Risk estimates were developed using benchmark dose methods applied to rodent dose-response data of adverse lung effects following subchronic or shortterm exposure to various types of SWCNT and MWCNT. In the absence of validated lung dosimetry models for CNT, lung doses were estimated assuming either deposited or retained lung dose in animals and humans. These findings suggest that workers are at risk of developing adverse lung effects, including pulmonary inflammation and fibrosis, if exposed to CNT over a working lifetime. Based on the two rat subchronic inhalation studies for two types of MWCNT (with different metal content), working lifetime exposures of 0.2–2 $\mu g/m^3$ (8-hr TWA; 95% LCL estimates) are estimated to be associated with a 10% excess risk of earlystage lung effects (minimal severity grade 1) (Table A–5). For a severity level of slight/mild (grade 2), the 45-year working lifetime excess risk estimates are approximately 0.7–19 $\mu g/m^3$ (8-hr TWA; 95% LCL estimates) (Table A–6).

These working liftetime 8-hr TWA concentrations are below the estimated upper LOQ (7 $\mu g/m^3$) of NIOSH Method 5040 for measuring the respirable mass concentration of CNT in air as an 8-hr TWA. Similar risk estimates relative to the LOQ were also derived for SWCNT and MWCNT from the shortterm studies, regardless of whether the CNT were purified or unpurified (with different types and amounts of metals), i.e., 0.08–12 $\mu g/m^3$ (Tables A–3 and

A–4). Lower risks are estimated at the lower LOQ of 1 μg/m^3, which are approximately 0.5% to 16% based on the rat subchronic dose-response data for the slight/mild lung effects and different lung dose estimation (95% UCL estimates) (Table A–8). Higher risks are estimated for the more sensitive endpoint of minimal grade 1 lung effects (Table A–7). Additional analyses and risk estimates based on other methods and assumptions are provided in Section A.6.

A.6. Sensitivity Analyses

Specific areas of uncertainty in this CNT risk assessment are evaluated in this section, including: (1) the rat lung dose estimation; (2) the critical effect level selection in animals and relevance to humans; and (3) alternative assumptions used in the OEL estimation methods. Sensitivity analyses in these areas were performed to qualitatively and quantitatively evaluate the influence of the different options and assumptions on the draft REL [NIOSH 2010].

A.6.1. Lung Dose Estimation

Key factors that influence the estimates of CNT lung burden in rats and humans include: (a) the lung geometry and airway dimensions; (b) lung and breathing parameters (including, functional residual capacity, total lung capacity, breathing frequency, and tidal volumes; (c) lung retention kinetics; and (d) interspecies dose normalization. The deposition fraction is based on the airborne particle size (and to some extent shape for nonspherical particles), on the breathing pattern (nasal, oral, or combination) and minute ventilation, and on the lung airway geometry. The ventilation rate depends on the species and on the activity level. Reference values are available for the average ventilation rates in rats and humans [EPA 1988, 1994; ICRP 1994]. The airborne particle size data (as reported in the animal studies) (Table A–2) were used to estimate the deposited lung dose of CNT in rats and humans, using spherical particle based models. The longterm clearance kinetics have been well studied and validated for inhaled poorly soluble spherical particles in rats [Anjilvel and Asgharian 1995; Asgharian et al. 2001, 2003] and in humans [ICRP 1994; Kuempel et al. 2001a, b; Gregoratto 2010, 2011], but models specifically for CNT are not yet available.

This section examines some of the key parameter values used in the lung dose estimation, and also characterizes the quantitative influence of alternative models and assumptions. Two studies were available to evaluate the lung dose estimates in rats. Pauluhn [2010a] and Ellinger-Ziegelbauer and Pauluhn [2009] provided cobalt tracer-based measurements of the CNT lung burden based on cobalt-tracer measurements. These data were used to compare MPPD model-based estimates. Because of prediction equation changes in the MPPD model from version 2.0 to 2.1, which affect the model-predicted rat alveolar deposition fraction predictions (discussed further in Section A.2.2), the cobalt tracer-based estimates are compared to each model version (Section A.6.1.2). The influence of assumed density on the CNT lung deposition fraction is quantified in addition to the evaluation of the MPPD model version 2.0 vs. 2.1 predictions (Section A.6.1.1). The derivation of allometric- based (body weight scaled) lung ventilation rate estimates is also discussed (Section A.6.1.3).

A.6.1.1. Lung Dosimetry Model- based Deposition Fraction and Dose Estimates

The fraction of inhaled CNT that is deposited in the respiratory tract is predicted from the aerosol characteristics. The deposition mechanisms include impaction, sedimentation, interception, and diffusion. The aerodynamic diameter, by definition, represents the gravitational settling (sedimentation) behavior of particles [Hinds 1999]. The definition of aerodynamic diameter standardizes the shape (to spherical) and density (to that of water, 1 g/ml). The aerodynamic diameter of a particle, regardless of its shape and density, is the diameter of a sphere with the same gravitational settling velocity as the particle in question. Conventionally, aerodynamic diameter has been used as a reference diameter to represent total particle deposition in the respiratory system over a wide particle size range. Models such as MPPD [CIIT and RIVM 2006; ARA 2011] use particle density (specified by the user), to convert aerodynamic to physical diameter and vice versa, and in this manner capture the key particle deposition mechanisms for spherical particles.

However, for high-aspect ratio particles and particles less than 500 nm diameter, including some individual or airborne agglomerates of CNT, the aerodynamic diameters are much smaller than their diffusion-equivalent diameter (i.e., the measure of diameter that captures the diffusional deposition mechanism) [Baron et al. 2006; Kulkarni et al. 2009]. When the different equivalent diameters could significantly differ, it is recommended to experimentally measure these property-equivalent diameters, and subsequently use the measured diameters in the lung deposition models to provide a reliable representation of each relevant deposition mechanism [Kulkarni et al. 2011].

In the animal inhalation studies of CNT [Shvedova et al. 2008; Ma-Hock et al. 2009; Pauluhn 2010a], the airborne particle sizes (MMAD) were in the micrometer size range (~1–3 µm) (Table A–2) and the airborne CNT structures in those studies were roughly spherical agglomerates—suggesting that deposition from diffusional mechanisms may be negligible and aerodynamic diameter may provide a reasonable estimate of the deposition efficiency of CNT in the respiratory tract. However, the density of the airborne structures can affect the deposition efficiency predictions in MPPD [ARA 2011]. An evaluation of the effect of the CNT density assumptions on the rat alveolar deposition fraction is provided in this section.

In the rat model, MPPD version 2.1 (but not 2.0) accepts density values less than one. The MMAD (GSD) values reported in the subchronic rat inhalation studies varied slightly with particle concentration and sampling device [Ma-Hock et al. 2009; Pauluhn 2010a]. The central MMAD (GSD) values were used for the deposition fraction and lung burden estimates. The influence of the alternative particle size estimates was not fully evaluated but appeared to be minimal compared with other factors (MPPD rat model version and assumed density).

In addition, the MPPD model estimates of CNT lung burden in rats are compared to the measured CNT lung burdens from two rat inhalation studies. Pauluhn [2010a] reported the amount of cobalt tracer in the rat lungs as well as the amount of Co that was matrix-bound to the CNT. The Ellinger-Ziegelbauer and Pauluhn [2009] 1-day inhalation study with 91-day post-exposure followup also reported Co data. These data provided a basis for comparison to lung burden estimates from the MPPD models.

Results in Table A–9 show that the rat deposition estimates (at the same density) vary by a factor of approximately two depending on the version of the MPPD model (2.0 or 2.1). As discussed in Section A.2.2, this is apparently because of a change in MPPD 2.1 in the deposition efficiency equations for the head region of the rat model, which reduces the deposition efficiency of the alveolar region. The lower density further reduces the alveolar deposition efficiency estimates. These findings suggest that rat alveolar lung dose estimates based on MPPD 2.1 (regardless of density assumption) would result in greater estimated potency of the CNT (because the response proportions do not change) and thus lower BMD(L) estimates in rats and lower OEL estimates (by approximately a factor of two) than those shown in the main analyses. Table A–9 also shows the human alveolar deposition fraction estimates from MPPD 2.0 and 2.1 (Yeh and Schum deposition model). MPPD 2.0 and 2.1 provide similar deposition fraction estimates for particle density of 1 g/ml. Different density assumptions (within MPPD 2.1) also had less effect (up to approximately 20%).

A.6.1.2. Cobalt Tracer vs. Dosimetry Model Estimates of MWCNT Lung Dose

Table A–10 provides a comparison of the dose estimates from either the MPPD 2.0 or 2.1 rat lung model to those from the cobalt tracer measurements reported in two studies [Pauluhn 2010a; Ellinger-Ziegelbauer and Pauluhn 2009].

Table A–10 shows that the cobalt-based estimate of CNT in the rat lungs is numerically between the deposited and retained dose estimated by MPPD 2.0 (density of 1). The MPPD 2.11 model (density of 0.2) [Pauluhn 2010b] underestimated the Co-based lung burden, even the deposited dose estimate (assuming no clearance). These findings suggest that the model-based estimates of the deposited and retained rat lung doses in the main analyses (MPPD 2.0, density 1) provided reasonable estimates of the bounds on the estimated lung burden. Moreover, these findings are consistent with the animal toxicokinetic data that show CNT overloads alveolar clearance at lower mass doses than for particles with lower total surface area or volume lung dose, resulting in increased retention of CNT in the lungs of rats and mice than expected for other poorly soluble respirable particles [Pauluhn 2010a; Mercer et al. 2009]. The finding that the cobalt-tracer estimates were between the deposited and retained lung doses is consistent with CNT reduced clearance compared with spherical particles.

Table A–9. Comparison of rat or human alveolar deposition fraction of inhaled particles, by MPPD version and density assumption*

Rat subchronic inhalation study	MPPD 2.0	MPPD 2.1	
	Density = 1 (g/ml)	Density = 1 (g/ml)	Density < 1 (g/ml)*
Rat estimates			
Ma-Hock et al. [2009]	0.072	0.044	0.024
Pauluhn [2010a]	0.046	0.027	0.023
Human estimates			
Ma-Hock et al. [2009]	0.099	0.10	0.080
Pauluhn [2010a]	0.086	0.090	0.084

* Density: 0.043 g/ml [Ma-Hock et al. 2009]; 0.2 g/ml [Pauluhn 2010a]. Particle MMAD (GSD): 1.2 (2.7) [Ma-Hock et al. 2009]; 2.74 (2.11) [Pauluhn 2010]; tidal volume 2.1 ml [Ma-Hock et al. 2009]; 2.45 ml [Pauluhn 2010a]; inhalability adjustment (all).

Table A–10. Comparison of MPPD model and cobalt-tracer based estimates of MWCNT lung burden—subchronic inhalation exposure in rats [Pauluhn 2010a]

Exposure concentration (mg/m^3)	MPPD 2.0*		MPPD 2.1†		MWCNT Retained lung dose (µg) estimated from cobalt-tracer‡
	Deposited lung dose (µg)	Retained lung dose (µg)	Deposited lung dose (µg)	Retained lung dose (µg)	
0.1	27	12	14	5.6	8.7
0.45	121	63	50	23	109
1.62	436	271	222	125	391
5.98	1,610	1,230	594	392	1,433

* MMAD (GSD)—2.74 (2.11); assumed density of 1 g/ml; tidal volume—2.45 ml; alveolar deposition fraction estimate was 0.046.

† Assumed density of 0.2 g/ml; tidal volume—2.45 ml; alveolar deposition fraction estimate was 0.023 for MMAD (GSD) of 2.74 (2.11).

‡ Mass of cobalt was estimated from Figure 6 in Pauluhn [2010a] to be approximately 10, 125, 450, and 1,650 ng, respectively, by increasing exposure concentration. The CNT amount in the lungs was estimated from the reported 0.115% Co that was matrix-bound to the CNT [Pauluhn 2010a]; the remaining mass (99.885% was assumed to be CNT). The CNT mass was thus calculated as CNT (ng) = [0.99885 × Co mass (ng)] / 0.00115. CNT (ng) × 0.001 µg/ng equals CNT (µg).

Table A–11. Comparison of MPPD model and cobalt-tracer based estimates of MWCNT lung burden—single day (6-hr) inhalation exposure in rats [Ellinger-Ziegelbauer and Pauluhn 2009]

Exposure concentration (mg/m^3)	MPPD 2.0*		MPPD 2.11†		MWCNT Retained lung dose (µg) estimated from cobalt-tracer‡
	Deposited lung dose (µg)	Retained lung dose (µg)	Deposited lung dose (µg)	Retained lung dose (µg)	
11	50	12	18	3.4	≤26
241	933	568	557	285	339

* MMAD (GSD)—2.9 (1.8) and 2.2 (2.6), respectively, for 11 and 241 mg/m^3, from Ellinger-Ziegelbauer and Pauluhn [2009]; alveolar deposition fraction—0.050 and 0.043, respectively, for 11 and 241 mg/m^3; assumed density of 1 g/ml; tidal volume—2.45 ml.

† Assumed density of 0.2 g/ml; tidal volume—2.45 ml; alveolar deposition fraction—0.019 and 0.026, respectively for 11 and 241 mg/m^3.

‡ Mass of cobalt at 91 d post-exposure was estimated from Figure 2 in Ellinger-Ziegelbauer and Pauluhn [2009] to be approximately 0.03 µg (11 mg/m^3) and 0.39 µg (241 mg/m^3). The CNT amount in the lungs was estimated from the reported 0.115% Co that was matrix-bound to the CNT [Pauluhn 2010a]; the remaining mass (99.885% was assumed to be CNT). The CNT mass was thus calculated as CNT (µg) = [0.99885 × Co mass (µg)] / 0.00115.

Similar comparisons were made of the cobalt-tracer or lung model estimated lung dose of MWCNT in a study of rats exposed for 1 day (Table A–11). Results show that the MPPD 2.0 model overestimated the retained lung dose of CNT by nearly a factor of two (at the higher dose) compared with the estimates based on the cobalt tracer in the Ellinger-Ziegel- bauer and Pauluhn [2009] study (Table A–11). This suggests greater clearance than would be predicted at this high dose (241 mg/m^3) based on overloading of lung clearance in the rat model (MPPD

2.0). If the retained lung dose estimated by cobalt tracer is the best estimate (closest to actual), this suggests that the BMD estimates using the model- estimated lung burdens may be overestimates (i.e., they underestimate potency because the response proportion is constant while the actual lung burden causing the effect may be lower). Some error may also exist in the cobalt-tracer measurements of the MWCNT mass (estimated from Figure 2 in Ellinger-Ziegelbauer and Pauluhn [2009]).

A.6.1.3. Pulmonary Ventilation Rate

The pulmonary ventilation rate influences the deposited and retained lung dose estimates. Rat inhalation rate estimates vary slightly among different sources [US EPA 1994; Pauluhn 2010a, citing Snipes 1989]. Species-specific ventilation rates can be calculated based on the following allometric scaling equation [US EPA 1994]:

Equation A.6:

$$\ln(V_E) = b_0 + b_1 \ln(BW)$$

where V_E is the minute ventilation (L/min); BW is body weight; and $b_0 + b_1$ are species-specific parameters.

For the rat, $b_0 + b_1$ are -0.578 and 0.821, respectively (Table 4–6 of US EPA [1994]). For a 300 g rat, the ventilation rate can be calculated as follows:

Equation A–7:

$$0.21 \text{ L/min} = \text{Exp}\,[-0.578 + 0.821 \times \text{Ln}(0.3)]$$

This is also the default minute ventilation in MPPD [CIIT and RIVM 2007; ARA 2011].

Rat mean body weights in Pauluhn [2010a] were reported as 369 g (male) and 245 g (female) in the control (unexposed) group at 13 weeks. Because the alveolar septal thickening response data in Pauluhn [2010a] were based on male rats only, a male rat minute ventilation of 0.25 L/min (calculated from equation A–6) was used to estimate lung dose in that study.

Ma-Hock et al. [2009] did not report the rat body weight, although the rat strain (Wistar) and study duration (13 weeks) were the same as in Pauluhn [2010a]. Because the granulomatous inflammation response data in Ma-Hock et al. [2009] were combined for the 10 male and 10 female rats in each dose group (since response proportions were statistically consistent), an average rat body weight in male and female rats of 300 g was assumed, which is similar to the male and female average body weight of 307 g reported in Pauluhn [2010a] and the default value of 300 g in MPPD. Subsequently, body weights were obtained for the Ma-Hock et al. [2009] study [personal communication, L. Ma-Hock and E. Kuempel, 10/14/10]. The average male and female rat body weight at 13 weeks was nearly identical (305 g) to that reported in Pauluhn [2010a]. Other rat minute ventilation rates of 0.8–1 L/min per kg [Pauluhn 2010a, citing Snipes 1989] would result in somewhat higher lung dose estimates.

Based on equation A–6, a minute ventilation of 0.21 L/min is calculated for female and male rats in Ma-Hock et al. [2009], and 0.25 L/min for male rats in Pauluhn [2010a]. Minute ventilation is the product of tidal volume and breathing frequency. Assuming the same breathing frequency (102 min^{-1}), a tidal volume of 2.45 ml is calculated (equation A–6) and used instead of the default value in MPPD 2.0 [CIIT and RIVM 2006] in estimating the rat lung dose in the Pauluhn [2010a] data.

In humans, based on the MPPD 2.0 model [CIIT and RIVM 2006], the default pulmonary ventilation rate is 7.5 L/min, based on default values of 12 min^{-1} breathing frequency and 625 ml tidal volume. The "reference worker" ventilation rate is 20 L/min [ICRP 1994] or 9.6 m^3/8 hr (given 0.001m^3/L, and 480 min/8-hr). In these estimates, 17.5 min^{-1} breathing frequency and 1143 ml tidal volume [NIOSH 2011a] were used in MPPD 2.0 to correspond to a 20 L/min reference-worker ventilation rate.

A.6.2. Critical Effect Level Selection

A key step in the dose-response analyses of any risk assessment is estimating the critical effect level. A critical effect level from an animal study is extrapolated to humans to derive a POD for low dose extrapolation (Section A.2.3). A critical effect is typically the most sensitive effect associated with exposure to the toxicant (i.e., the effect observed at the lowest dose) which is adverse or is causally linked to an adverse effect. The earlystage lung effects discussed in Section A.2.1.3 are the critical effects used in both the main risk assessment and in these sensitivity analyses. The primary critical effect selected is the proportion of rats with minimal (grade 1) or higher severity of pulmonary inflammation or alveolar septal thickening (as reported by Ma-Hock et al. [2009] and Pauluhn [2010a]). In addition, grade 2 (slight/mild) or greater effects (as reported in the same studies) were also evaluated as a response endpoint since the interpretation of the histopathology, for a slight or mild response, may be less variable than that for a minimal response, and may also be more relevant to a potential adverse health effect in humans.

Table A–12. Effect level estimates in rats after subchronic (13-wk) inhalation exposure to multiwall carbon nanotubes (MWCNT)

Study	Effect level in rats				BMR
	LOAEL (mg/m^3)	NOAEL (mg/m^3)	BMC (mg/m^3)	BMCL (mg/m^3)	
Ma-Hock et al. [2009]	0.1	nd	0.060	0.023	Granulomatous inflammation (≥ minimum, severity grade 1+)
	0.5	0.1	0.12	0.082	Granulomatous inflammation (≥ mild, severity grade 2+)*
Pauluhn et al. [2010]	0.45	0.1	0.10	0.051	Alveolar septal thickening (≥ minimal, severity grade 1+)
	1.5	0.45	0.87	0.45	Alveolar septal thickening (≥ mild, severity grade 2+)

Abbreviations: NOAEL: No observed adverse effect level; LOAEL: Lowest observed adverse effect level; BMC: Benchmark concentration (maximum likelihood estimate) associated with 10% excess risk of specified BMR. BMCL: 95% lower confidence limit of the BMC; based on a multistage model, polynomial degree 2, P = 0.88); BMR: Benchmark response; nd: not determined.

* Same response proportion per dose, and therefore the same BMD(L) estimates, for alveolar lipoproteinosis.

The critical effect levels in the main analysis are the BMD(L) estimates from the dose-response modeling of the rat estimated deposited or retained lung dose and to the human-equivalent lung dose estimates. The working lifetime exposure concentration that would result in that equivalent lung dose was then calculated, assuming either particle-size specific lung deposition only (assuming no clearance) or the estimated retained lung dose (assuming normal spherical particle clearance).

In the main risk analysis, BMD methods were selected over NOAELs or LOAELs because of several statistical advantages (Section A.2). However, BMD(L) estimates may also be uncertain, for example, when the dose spacing is not optimal, as occurred in the CNT subchronic studies (Figures A–1 and A–4). In this sensitivity analysis, NOAELs and LOAELs reported in the subchronic inhalation studies [Ma-Hock et al. 2009; Pauluhn 2010a] are used as the effect levels in evaluations of alternative methods to derive OEL estimates. A quantitative comparison of possible critical effect levels is shown in Table A–12. The BMDL estimates are generally similar to the NOAEL estimates (within a factor of approximately 1 to 4), which suggests that the BMDL estimates may be reasonable despite the sparse data in the low dose region of the subchronic inhalation studies (Figure A–1).

A statistical analysis was performed to compare the NOAEL and BMD estimates (in this example, the BMD is an exposure concentration, or BMC). The maximum likelihood estimate of the excess risk (of a minimal or higher grade of alveolar septal thickening) at 0.1 mg/m^3 is 0.10 (i.e., 10%), based on the BMD model fitted to the dose-response data in the Pauluhn [2010a] study (Table A–12). Yet, 0.1 mg/m^3 was identified as a NOAEL based on zero adverse response being observed [Pauluhn 2010a]. In order to assess the precision of the estimate of the excess risk associated with this NOAEL, the likelihood of the data in the NOAEL and control groups was reparameterized in terms of the respective sum and difference of the expected response proportions; and an upper confidence limit for the difference was assessed by inverting its likelihood ratio test statistic. When a nominal confidence coefficient of 95% for a two-sided interval was applied, a value of 0.17 (i.e., 17%) was obtained for the UCL of the difference. Hence, the results supporting the use of 0.1 mg/m^3 as a NOAEL are also statistically consistent with the results from the BMD model since the MLE of excess risk based on the model is less than the UCL.

In a standard risk assessment approach, BMDL estimates may be considered equivalent to a NOAEL for use as a POD in risk assessment [US EPA 1994]. Once an effect level is selected in a given animal study, it is extrapolated to a human-equivalent effect level (e.g., as 8-hr TWA concentration), or human-equivalent concentration (HEC). This HEC_POD (human-equivalent point-of-departure) is the POD for either extrapolating to a lower (acceptable) risk level or applying uncertainty factors in the derivation of an OEL. These steps are discussed further in Section A.6.3.

A.6.3. Alternative OEL Estimation Methods

As mentioned in the previous section, a standard risk assessment method using animal data typically involves first identifying a critical effect level in animals (e.g., NOAEL or BMDL), which is the POD_{animal}. A HEC_POD is estimated by extrapolating the animal dose to humans by accounting for the biological and physical factors that influence the lung dose across species[†]. Lung dosimetry models can account for these interspecies differences and provide equivalent dose estimates in animals and humans given the exposure concentration and duration, the breathing rates and patterns, and the physical properties of the aerosol. A simplified standard approach in lieu of a lung dosimetry model to apply a total dosimetric adjustment factor to the animal effect level (Section A.6.3.1). It is useful to evaluate both approaches given that the lung dosimetry models have not been specifically validated for respirable CNT.

A.6.3.1. Illustration of Human-Equivalent Concentration Estimation

The human equivalent concentration HEC) to a POD_{animal} (e.g., NOAEL) in an animal study can be calculated as:

Equation A–8:

$$HEC\_POD = POD_{animal} / DAF$$

where DAF is the dosimetric adjustment factor, and

Equation A–9:

$$DAF = (VE_H/VE_R) \times (DF_H/DF_A) \times (RT_H/RT_A) \times (NF_A/NF_H)$$

where VE is the ventilation rate (e.g., as total volume of air inhaled per exposure day, m^3/d) in humans (H) or animals (A); DF is the deposition fraction, in this case, in the alveolar region of the respiratory tract; RT is retention half-time of particles in the lungs, and NF is the interspecies dose normalization factor.

The basic method shown in equation A–9 is consistent with the reference concentration (RfC) method [US EPA 1994] and the method used by Pauluhn [2010b].[‡] The NOAEL of 0.1 mg/m^3 and the specific adjustment factors reported by Pauluhn [2010b] are used below as an example of this standard risk assessment approach to estimating a HEC_POD:

Equation A–10:

$$AF^{\ddagger} = (0.14/0.29) \times (0.118/0.057) \times (10/1) \times (8.66 \times 10^{10}/4.99 \times 10^{11}) = 1.73$$

VE (Equations A–9 and A–10) is expressed as volume per body weight (m^3/kg) Pauluhn [2010b]. The rat value of 0.29 is calculated from a ventilation rate of 0.8 L/min/kg BW × 360 min (i.e., rat 6-hr exposure day) × 0.001 m^3/L. The human value of 0.14 is based on 9.6 m^3 (per 8-hr workday)/70 kg. The alveolar DF is for a MMAD of 3 μm and estimated from the

[†] HEC_POD is then divided by appropriate uncertainty factors (UFs) to account for variability and uncertainty in its estimation (Section A.6.3.3).

[‡] There are some differences in the terminology and presentation of standard methods to estimate a HEC_POD. For example, the placement of the animal or human factors in the numerator or denominator of the ratios determines whether the effect level is multiplied or divided by the DAF). The term AF (adjustment factor [Pauluhn 2010b]) is equivalent in concept to the term DAF [US EPA 1994]. AF is used here to clarify that the adjustment factor illustrated here is specific to the estimation of the human-equivalent dose from the Pauluhn [2010b] study.

MPPD2.0 rat and human models [CIIT and RIVM 2006]. RT is based on an assumed average retention half-time of approximately 1–2 yr in humans and 60 days in rats [Snipes et al. 1989] (the factor was rounded to 10 in Pauluhn 2010b). NF is expressed as the total alveolar macrophage cell volume (μm^3) per kg BW in each species, calculated from the average AM cell volume (1166 μm^3 rat, 4990 μm^3 human), the total average number of AMs per lung (2.6×10^7 rat, 7.0×10^9 human), and the average BW in rats and humans (0.35 and 70 kg, respectively) [Pauluhn 2010b] (Table A–2).

The AF of 1.7 was rounded up to 2 in Pauluhn [2010b]. Thus,

Equation A–11:

$$HEC\_POD = 0.1 mg/m^3 / 2 = 0.05\ mg/m^3$$

The method shown in equations A–8 to A–11 is seen to be identical to the derivation of the humanequivalent NOAEL in Pauluhn [2010b]. In that study, the HEC_NOAEL was not divided by UFs, but was proposed as an OEL [Pauluhn 2010b].

To simplify further, the BW factors in equation A–10 can be dropped. The BW terms cancel out such that the estimate is the same whether or not BW is used in these calculations. Thus, substituting the terms expressed per kg BW in equation A–10 with the animal or human values of VE_H/VE_R (9.6/0.1015) and NF_A/NF_H ($3.03 \times 10^{10} / 3.49 \times 10^{13}$) results in the same AF of 1.7.

A.6.3.2. Evaluation of Dosimetric Adjustment Factors

As illustrated in Section A.6.3.1 and equation A–9, the four adjustment factors that make up the total DAF include:

1. Air intake (ventilation rate)
2. Deposition fraction
3. Dose retention
4. Normalizing factor

To evaluate the quantitative effect of the different assumptions in extrapolating the animal dose to humans, it is of interest to examine alternative assumptions in these four adjustment factors (Equation A–9). The first two factors—ventilation rate (VE) and deposition fraction (DF)—do not vary much among the various sources because they are based on well-known physiological and physical measurements and models. There is some uncertainty in the DF estimates for CNT as they are based on models for spherical particle aerosols, and also the density assumption can influence the DF estimate, although a larger difference (approximately a factor of two) is due to differences in the spherical particle model-based estimates (e.g., MPPD 2.0 vs. 2.1) (Sections A.2.2; A.6.1.1; and A.6.1.2).

The values used by NIOSH [2010] and Pauluhn [2010b] are similar for the VE and DF, i.e., for human and rat, respectively:

- VE (m^3/d): 10 and 0.015 [Pauluhn 2010b], 9.6 and 0.09 [NIOSH 2010]; and
- DF: 0.118 and 0.057 [Pauluhn 2010b], 0.086 and 0.046 [NIOSH 2010].

The other two factors—retention half-time (RT) and interspecies normalization factor (NF)—can differ largely depending on the assumed mode of action concerning how the deposited CNT interacts with the lung tissue over time. These factors are discussed below.

A.6.3.2.1. Interspecies Dose Normalization Factor

The interspecies NF adjusts for the size difference in the lung (surface area or volume) into which the CNT dose deposits. Studies of other inhaled particles or fibers are relevant to evaluating mechanisms that may also apply to CNT in the lungs. Possible dose metrics related to the modes of action for pulmonary inflammation and fibrosis include the CNT mass, surface area, or volume dose per alveolar epithelial cell surface area or alveolar macrophage cell volume in each species. Normalizing the dose (e.g., NOAEL) across species to the total average alveolar macrophage cell volume in rat or human lungs is based on the experimental observation of overloading of alveolar clearance in rats and mice exposed to respirable poorly soluble particles or fibers [Bolton et al. 1983; Morrow 1988; Bellmann et al. 1991; Elder et al. 2005; Pauluhn 2010b].

(a) Alveolar macrophage cell volume

At a sufficiently high particle dose, pulmonary clearance can become impaired due to overloading of alveolar macrophage-mediated clearance. In rats, the overloading dose has been observed as particle mass (~1 mg/g lung), volume (~1 μl/g lung for unit density particles) [Morrow 1988; Muhle et al. 1990], or surface area (200–300 cm^2 particles per rat lung) [Tran et al. 2000]. On a volume basis, an overloading particle dose corresponds to approximately 6%–60% of total alveolar cell volume, when overloading begins and is complete, respectively [Morrow 1988]. The 60% value has been observed experimentally [Oberdörster et al. 1992], although particle clearance impairment may start at lower particle volume lung dose [Bellmann et al. 1991; Kuempel et al. 2001a]. Biological responses to overloading include: accumulation of particle-filled macrophages in the alveoli, increased permeability of the epithelial cell barrier, persistent inflammation, increased particle translocation to the alveolar interstitium and lung-associated lymph nodes, as well as increasing alveolar septal thickening, lipoproteinosis, impaired lung function, and fibrosis [Muhle et al. 1990, 1991].

Although the overload mode of action in the rat has been well-studied, the extent to which overloading is involved in human lung responses to inhaled particles is not as clear due to observed differences in both the kinetics and the pattern of particle retention in the lungs of rats and humans. Whereas particle clearance in rats is first-order at doses below overloading, studies in workers have shown that human lung clearance of respirable particles is not first-order even at relatively low retained particle mass lung low doses [Kuempel 2000; Kuempel et al. 2001; Tran and Buchanan 2000; Gregoratto et al. 2010, 2011]. That is, some portion of the particle dose that deposits in the pulmonary region is retained for a very long time (retention half-time of several years) [ICRP 1994; Kuempel et al. 2001; Gregaratto et al. 2010]. Humans also apparently retain a greater portion of the particles in the alveolar interstitium, whereas rats retain more particles in the alveolar space [Nikula et al. 1997, 2001]. The greater interstitial particle retention may increase the dose to the target tissue for pulmonary fibrosis in humans relative to that for the same deposited dose in rats lungs. Given the differences in the particle clearance kinetics and retention patterns in rats and humans, normalizing the dose across species based on the total alveolar macrophage volume may not be the best dose metric for predicting adverse lung responses in humans.

(b) Alveolar epithelial cell surface area

Another dose metric that may be relevant to the inflammatory and fibrotic lung responses is the particle or CNT dose per surface area of alveolar epithelial cells [US EPA 1994; Donaldson et al. 2008]. It is the epithelial cell surface with which particles interact when they migrate through the epithelial cell layer into the interstitium, and epithelial cells are also involved in recruitment of inflammatory and fibrotic cells [Bohning and Lippmann 1992; Driscoll et al. 1996; Tran et al. 2000]. For this reason, normalizing the dose based on the total alveolar epithelial cell surface area may be more predictive of the human lung response. However, since both the alveolar macrophages and epithelial cells are involved in the lung responses to inhaled particles, some combination of dose metrics may ultimately be most predictive in this dynamic biological system.

In the absence of a more complete biologically-based model, an evaluation of the quantitative influence of each assumed dosimetric mode of action (e.g., based on either the alveolar macrophage cell volume or the epithelial cell surface area) provides information on the sensitivity of the risk assessment and OEL derivation to the interspecies dose normalization factor. Thus, replacing the alveolar macrophage volume ratio in equation A–10 with a NF_A/NF_R of $0.4m^2/102m^2$ [Stone et al. 1992] results in a total AF that is 4.5 × greater. That is,

Equation A–12:

$$AF = (9.6m^3/0.102m^3) \times (0.118/0.057) \times (10/1) \times (0.4m^2/102m^2) = 7.7$$

Equation A–13:

$$HEC\_NOAEL = 0.1mg/m^3 / 7.7 = 0.013 \ mg/m^3$$

The larger AF results in a corresponding smaller human-equivalent concentration. This illustrates that the risk estimates for CNT—as for other inhaled particles—is sensitive to the assumed mode of action concerning the interspecies normalizing factor.

A.6.3.2.2. Interspecies Dose Retention Factor

The retained dose to the target tissue is influenced by the clearance mechanism in the lung region in which the particles deposit. RT in equation 2 (as the kinetic factor in Pauluhn [2010b]) is intended to account for the differences in the rat and human particle retention half-time. This factor is also dependent on the assumptions concerning the biological mode of action. In the rat, evidence suggests that doses of poorly soluble low toxicity particles below those causing overloading of lung clearance (i.e., at steady-state) would not be associated with adverse lung effects. A steady-state lung burden means that the rate of particle deposition equals the rate of clearance such that once the steady-state burden had been achieved, the lung burden would be the same over time if exposure conditions did not change. For example, if steady-state lung burden was reached after subchronic (13 week) exposure to a given exposure concentration, then the chronic (2 yr) lung burden would be the same given the same rates of exposure and clearance. However, the steady-state lung burden may not been entirely reached by 13 weeks in the rat or in an equivalent time in humans. Based on the rat overload mode of action, Pauluhn [2010b] assumed that humans would achieve a steady-state lung burden if exposed at an equivalent total particle volume dose in the alveolar macrophages (over a roughly equivalent human exposure duration of 10 years to a rat 3 month exposure). A ratio of 10/1 for human/rat retention half-time rate was used [Pauluhn 2010b], based on a simple first-order clearance rate model of particle clearance from the lungs in both rats and humans [Snipes et al. 1989]. The volumetric dose of CNT associated with overloading in the rat was equivalent to a relatively low mass dose compared to other poorly soluble particles [Pauluhn 2010b. 2011]. Moreover, human lung-particle retention data have shown that a simple first-order clearance model would underpredict the human longterm lung dose at similar low mass doses [ICRP 1994; Kuempel et al. 2001; Gregoratto et al. 2010]. That is, the human longterm retained lung burden would be expected to exceed a steady-state lung burden predicted from the rat model (i.e., low-dose first-order clearance with dose-dependent impairment, or overloading, of particle clearance after reaching a critical lung dose).

An alternative approach evaluated was to use the MPPD 2.0 [CIIT and RIVM 2006] human lung dosimetry model to estimate directly the retained lung burden in humans over a working lifetime. This approach assumes a mode of action in which the cumulative retained particle dose is related to the adverse lung responses, regardless of the dose rate (i.e., the time required to reach that dose). The cumulative exposure concept (concentration × time), known as “Haber’s Law,” is a typical default assumption in risk assessment for longterm exposures in the absence of other data [US EPA 1994]. Some studies in workers (coal miners) have shown that the working lifetime cumulative exposure and the retained lung dose are better predictors of pulmonary fibrosis than the average exposure concentration without consideration of duration [NIOSH 2011b]. Yet, there remains uncertainty about how well a cumulative dose received over a short duration may predict the response to the same cumulative dose received over a longer duration (i.e., at a lower dose rate). The direction of error could go either way, depending on the biological mechanisms of response. For example, a lower dose rate may allow the lung defense mechanisms to adapt to the exposure (e.g., by increasing clearance or repair mechanisms), which could reduce the adverse response at a later time point. On the other hand, a longer time in which a substance is in contact with the tissue may exacerbate the response, resulting in a more severe effect at the later time point. The actual lung response may be some combination of these effects.

To evaluate the assumptions used to estimate the human and rat retention kinetics, estimates were compared from the MPPD2.0 lung dosimetry model [CIIT and RIVM 2006] to the ratio of 10/1 for RT_H/RT_A used by Pauluhn [2010a]. The rat and human lung dosimetry models take into account the ventilation rates, the deposition fraction by respiratory tract region (predicted from particle size and breathing rate and pattern, nasal vs. oronasal), and the normal average clearance rates. Using the particle size and breathing rate values for the Pauluhn 2010a study (Table A–2), the rat retained lung burden at the end of the 13 week exposure to 0.1 mg/m^3 was estimated to be ~12 µg (Table A–10). This is similar to the 8.7 µg lung burden estimated from the cobalttracer based measurement (Table A–10).

Assuming that the rat has achieved steady-state lung burden after 13 wk exposure to 0.1 mg/m^3, the chronic lung burden should also be approximately 12 µg.[§] Extrapolating the rat lung dose of 0.012 mg to the human-equivalent lung burden would result in either:

- 13.5 mg—estimated by dividing the rat lung dose by an interspecies NF for the average total alveolar macrophage cell volume (i.e., 3.03×10^{10} µm^3/3.49×10^{13} µm^3) (rat/human); or
- 3.0 mg based on the average total alveolar epithelial cell surface area (0.4 m^2/102 m^2) (rat/human).

The associated 8-hr TWA concentration for 45-yrs would result in human-equivalent lung burdens (estimated from MPPD 2.0 human model [CIIT and RIVM 2006]) of 16 µg/m^3 and 3.5 µg/m^3, respectively, for the normalized lung burdens based on the alveolar macrophage cell volume or the alveolar epithelial cell surface area (Table A–13). The value of 16 µg/m^3 is approximately 3-fold lower than the ~50 µg/m^3 as the human equivalent concentration to the rat NOAEL reported in Pauluhn [2010b] (or 3.5 × lower than the 58 µg/m^3 HEC_LOAEL by applying an AF of 1.7 without rounding to 2). This difference is due to the approximately 3x higher retained lung dose estimates after a 45-year working lifetime (Table A–14) to that estimated as a 10-year steady-state lung burden [Pauluhn 2010a]. This suggests that the RT of 10 may underestimate the rat to human lung retention kinetics, and that a factor of 35 (i.e., 10 × 3.5) may be more realistic. Since the MPPD model already takes into account the ventilation rate and deposition fraction, the difference in the human retained lung dose estimates is due to greater particle retention predicted by the MPPD model (which includes the ICRP [1994] clearance model) [CIIT and RIVM 2006] compared to that of the first-order kinetic model used to estimate the factor of 10/1 [Snipes et al. 1989; Pauluhn 2010b].

When this same method was applied to the rat LOAEL of 0.1 mg/m^3 in the Ma-Hock et al. [2009] subchronic inhalation study, but using the particle size data and rat minute ventilation specific for that study (Table A–2 and Section A.2.2), similar human-equivalent estimates were obtained. The slightly higher doses are due to the greater DF for the MWCNT in the Ma-Hock et al. [2009] study. In addition, the POD from the Ma-Hock et al. [2009] study is based on a LOAEL (vs. NOAEL in Pauluhn [2010a], so an additional uncertainty factor would be applied (as discussed in the next section). In each case, the estimates using an interspecies normalizing factor based on the alveolar epithelial cell surface area are lower by a factor of approximately four. The estimates in Table A–13 are working life-time human-equivalent concentrations to the rat NOAEL or LOAEL with no uncertainty factors applied to these values. Lower estimated rat lung doses and human-equivalent lung doses and associated working lifetime concentrations would be expected if using MPPD 2.1 and density <1 g/ml (Section A.6.1.1 and Table A–9).

A.6.3.3. Selection of Uncertainty Factors and OEL Derivation

Standard noncancer risk assessment often assumes a threshold model, such that exposures below the OEL are assumed to be associated with essentially zero risk. Uncertainty factors

---

[§] At 2 years, the MPPD model predicted a lung burden of 13 µg and a lung burden plus lung-associated lymph node burden of 23 µg.

(UFs) are applied to the POD estimates (e.g., HEC_NOAEL) to derive health-based OELs. Uncertainty factors include not only factors to account for the uncertainty in the models and estimates used in the risk assessment, but also to account for variability in the distribution of responses in the human population. Possible UFs are shown in Table A–14. Various criteria and systems for UF selection have been developed (e.g., US EPA [1994]; WHO [2005]). NIOSH has used these standard systems to select uncertainty factors in the derivation of RELs on a case by case basis. The selection of UFs for a given study may depend in part on the criteria or system used and the interpretation of the available dose and response data within that system. Other risk assessments of CNT that have used uncertainty factors include Aschberger et al. [2010] and Nakanishi [2011a].

To obtain an OEL estimate based on the rat NOAEL or LOAEL data, the human-equivalent NOAEL or LOAEL (e.g., Table A–13) would be divided by study or data-based UFs (e.g., Table A–14). Using this approach, the human-equivalent NOAEL or LOAEL estimates of 3.5 to 18 μg/m^3 (Table A–13) divided by example uncertainty factors of 20 or 60 (Table A–14) results in estimates of approximately <1 μg/m^3 as the working lifetime exposures likely to be without appreciable risk of adverse effects. Such OEL estimates are below the upper LOQ (7 μg/m^3) of the analytical method to measure airborne CNT [NIOSH method 50540] and approximately equal to or less than the lower LOQ of 1 μg/m^3 [NIOSH method 5040]. These findings are consistent with the BMD-based estimates, which generally indicate >10% excess risk at the LOQ, depending on the effect level (BMR of grade 1+ or grade 2+ lung effects) and lung dose (deposited or retained) assumptions (Tables A–3 through A–6). That is, the working lifetime 10% BMC(L) estimates are generally less than 1 or 7 μg/m^3. Other estimates indicate <10% excess risk at the lower LOQ of 1 μg/m^3, based on an effect level (BMR) of histopathology grade 2 or higher. At the REL of 1 μg/m^3, the 45-yr working lifetime risk estimates of slight or mild (grade 2) lung effects, based on the rat subchronic inhalation studies to MWCNT, were approximately 0.5 to 16% (8-hr TWA; 95% UCL estimates) (Table A–8), depending on the type of MWCNT and the estimated lung dose in animals and humans.

Excess risk estimates based on the shortterm studies of SWCNT and MWCNT in rats and mice (Tables A–3 and A–4) are consistent with those from the rat subchronic inhalation studies (Tables A–5 and A–6). However, the uncertainty factors applied to the shortterm studies would be expected to be higher (e.g., by factor of 2) [Nakanishi 2011a] than those for the subchronic studies.

A.6.4. Summary of Sensitivity Analyses Findings

Many of the areas of uncertainty in these risk estimates for CNT also occur in other standard risk assessments based on subchronic or shortterm animal study data. Potential chronic effects of CNT are an important area of uncertainty because no chronic study results were available. Uncertainty exists about the estimated lung doses for the inhalation studies because of lack of experimental evaluation or validation of lung dosimetry models to predict deposition and retention of CNT. Information is also limited on the relative potency of different types of CNT to cause specific lung effects in animals because of study differences. Despite the variability in the risk estimates across the various types of CNT, all of the risk estimates were associated with low-mass concentrations (below the upper and lower LOQ or 7 or 1 μg/m^3, respectively).

Table A–13. Human-equivalent retained lung burden and working lifetime 8-hr TWA concentrations to rat subchronic NOAEL or LOAEL of 0.1 mg/m³

Subchronic rat study and normalizing factor	Rat lung burden (μg)†	Human-equivalent lung burden (mg)‡	Working lifetime 8-hr TWA (μg/m³)§
Pauluhn [2010a]			
Alveolar macrophage volume	11.7	13.5	16
Alveolar epithelial cell surface area		3.0	3.5
Ma-Hock et al. [2009]			
Alveolar macrophage volume	16.0	18	18
Alveolar epithelial cell surface area		4.1	4.0

† Estimated retained lung burdens in rats from MPPD 2.0 [CIIT and RIVM 2006], assuming particle size (MMAD and GSD) in Table A–2 and unit density. MPPD adjusts for the rat ventilation rate, deposition fraction by respiratory tract region, and uses a first-order clearance model (with rat overload at higher doses).

‡ Human-equivalent retained lung burden estimated by dividing the rat lung burden by a normalizing factor for interspecies differences—either the total alveolar macrophage cell volume (3.03×10^{10} μm^3/3.49×10^{13} μm^3) (rat/human) or the total alveolar epithelial cell surface area (0.4 m^2/102 m^2) (rat/human).

§ Human-equivalent concentration and point of departure (HEC_POD) based on the rat NOAEL [Pauluhn 2010a] or LOAEL [Ma-Hock et al. 2009], as 8-hr TWA over a 45-year working lifetime, estimated using MPPD 2.0 [CIIT and RIVM 2006] (Yeh and Schum human deposition model), reference worker ventilation rate and patterns [ICRP 1994], and the same particle size in Table A–2.

In conclusion, these sensitivity analyses show that the estimates of a health-based OEL are not strongly dependent on the BMD-based risk assessment methods, and the use of an alternative (POD/UF) method provides supporting evidence indicating the need for a high level of exposure containment and control for all types of CNT.

A.7. Evaluation of Carbon Nanofiber Studies in Mice and Rats

Two *in vivo* studies of carbon nanofibers (CNF) have been recently published in mice and rats [Murray et al. 2012 and DeLorme et al. 2012]. In order to compare the lung responses to CNF observed in these studies, estimates of the lung doses normalized across species are provided in this section.

A.7.1. Particle Characteristics

Both types of CNF were vapor grown, but obtained from different sources. In Murray et al. [2012], the CNF was supplied by Pyrograf Products, Inc. The chemical composition was 98.6% wt. elemental carbon and 1.4% wt. iron. CNF structures were 80 to 160 nm in diameter, and 5 to 30 μm in length. The specific surface area (SSA) measured by BET was 35-45 m^2/g; the effective SSA was estimated as ~21 m^2/g [Murray et al. 2012]. In DeLorme et al. [2012], the CNF was supplied by Showa Denko KK, Tokyo, Japan. The chemical composition was >99.5% carbon, 0.03% oxygen and < 0.003% iron. CNF structures were 40-

350 nm (158 nm average) in diameter and 1-14 μm in length (5.8 μm average). The BET SSA was 13.8 m^2/g [DeLorme et al. 2012].

A.7.2. Experimental Design and Animals

The species and route of exposure also differed in the two studies. In Murray et al. [2012], six female C57BL/6 mice (8-10 wk of age, 20.0 ± 1.9 g body weight) were administered a single dose (120 μg) of CNF by pharyngeal aspiration [Murray et al. 2012]; mice were examined at 1, 7, and 28 days post-exposure. In DeLorme et al. [2012], female and male Crl:CD Sprague Dawley rats (5 wk of age) were exposed to CNF by nose-only inhalation at exposure concentrations of 0, 0.54, 2.5, or 25 mg/m^3 (6 hr/d, 5 d/wk, 13 wk). The rats were examined 1-d after the end of the 13-wk exposure and 3 months post-exposure. Body weights were reported as: 252 g ± 21.2 female; 520 g ± 63.6 male (unexposed controls, 1-d post-exposure); 329 g ± 42.2 female; 684 g ± 45.8 male (unexposed controls, 3 mo. post-exposure) [DeLorme et al. 2012].

Table A–14. Example of uncertainty factor (UF) selection for human-equivalent concentrations to rat effect level in MWCNT subchronic inhalation studies†

Type of UF	Possible UF value	NOAEL [Pauluhn 2010a]	LOAEL [Ma-Hock et al. 2009]	Rationale
1. Animal to human extrapolation	Up to 10 (4 for TK; 2.5 for TD) [WHO 2005]	2	2	TK: Dosimetry based on estimated retained lung burden (in Table A–13); additional uncertainty about slower clearance of CNT (ad hoc TK factor of 2). TD: Assume equal average subchronic response at equivalent dose in each species (TD factor of 1).
2. Sub-chronic animal dose-response data used in absence of chronic data	Up to a factor of 10 [US EPA 1994]	2	2	Subchronic data were used, assuming steady-state lung burden in rats uncertainty about chronic effects (ad hoc UF of 2).
3. Human inter-individual variability for ensitive sub-population	10 (3.16 each for TK and TD) [WHO 2005]	5	5	Variability in workers (TK and TD components); factor of 5 from Aschberger et al. [2010].
4. LOAEL used instead of NOAEL	Up to a factor of 10 [US EPA 1994]	1	3	NOAEL: factor of 1. LOAEL: Responses at LOAEL were "minimal" severity by istopathology [Ma-hock et al. 2009].
5. Modifying factor (e.g., poor data quality; severe effect)	Up to a factor of 10 [US EPA 1994]	1	1	Subchronic studies were standard quality assays, and lung effects were earlystage.
Total UF	3,000‡	20	60	OEL is derived as HEC_POD/ total UF.

Abbreviations: TK—toxicokinetic. TD—toxicodynamic. NOAEL—no observed adverse effect level. LOAEL—lowest observed adverse effect level. POD—point of departure. UF—uncertainty factor.

† These UF examples refer to estimates in Table A–13.

‡ Total uncertainty factor is typically capped at 3,000 [US EPA 1994].

A.7.3. Lung Responses

In mice, the lung responses to CNF included pulmonary inflammation (polymorphonuclear lymphocytes, PMNs, measured in bronchioalveolar lavage fluid, BALF); PMN accumulation in CNF- exposed mice was 150-fold vs. controls on day 1. By day 28 post-exposure, PMNs in BALF of CNF- exposed mice had decreased to 25-fold vs. controls. Additional lung effects included increased lung permeability (elevated total protein in BALF), cytotoxicity (elevated lactate dehydrogenase, LDH), which remained significantly elevated compared to controls at day 28 post-exposure. Oxidative damage (elevated 4-hydroxynonenol, 4-HNE, and oxidatively modified proteins, i.e., protein carbonyls) was significantly elevated at days 1 and 7, but not at day 28. Collagen accumulation at day 28 post- exposure was 3-fold higher in CNF-exposed mice vs. controls by biochemical measurements. Consistent with the biochemical changes, morphometric measurement of Sirius redpositive type I and III collagen in alveolar walls (septa) was significantly greater than controls at day 28 post-exposure [Murray et al. 2012].

In rats, the respiratory effects observed in the De-Lorme et al. study [2012] were qualitatively similar to those found in the Murray et al study [2012]. The wet lung weights were significantly elevated compared to controls in male rats at 25 mg/m^3 CNF and in female rats at 2.5 and 25 mg/m^3 CNF at 1-day post-exposure; lung weights remained elevated in each sex in the 25 mg/m^3 exposure group at 3 mo. post-exposure. Histopathologic changes at 1 day post-exposure included inflammation in the terminal bronchiolar and alveolar duct region in the 2.5 and 25 mg/m^3 exposure groups, and interstitial thickening with type II pneumocyte proliferation in the 25 mg/m^3 exposure group. Cell proliferation assays confirmed increased cell proliferation in that highest dose group in the subpleural, parenchymal and terminal bronchiolar region; the subpleural proliferation in this dose group did not resolve in the females by the end of the 3 month recovery period. Cell proliferation appeared to resolve in males after a 3 month recovery period but numerically remained higher in the parenchyma and subpleural regions. Histopathologic evidence of inflammation and the presence of fiber-laden macrophages were reported to be reduced but still present in the high dose group after a 3 month recovery period. Inflammation within the alveolar space (as measured by PMN levels in BALF) was statistically significant only in the rats exposed to 25 mg/m^3 CNF. However, the percent PMNs increased in a dose-responsive manner: 1.2 (± 0.81), 1.4 (± 0.79), 2.7 (± 0.67), and 11 (± 2.0), respectively, in the 0, 0.54, 2.5, and 25 mg/m^3 exposure groups. LDH and other BALF markers were elevated at the end of the 13- wk exposure only in the 25 mg/m^3 exposure group, and LDH remained elevated at 3-mo. post-exposure in that group. The observed no adverse effect level (NOAEL) in rats was reported to be 0.54 mg/m^3. The lowest observed adverse effect level (LOAEL) was reported to be 2.5 mg/m^3 "...based on the minimal inflammation observed in terminal bronchioles and alveolar ducts of male and female rats..." [De-Lorme et al. 2012].

The sample size and sensitivity of the markers or assays are factors that could influence the statistical power and likelihood of observing exposure-related effects in these animal studies. In Murray et al. [2012], six animals per group were used for the BAL analysis, histopathology evaluation, oxidative stress markers, and lung collagen measurements. Five animals per group were used for the BAL and cell proliferation assays in the DeLorme et al. [2012] study (male and female data were analyzed separately). The Murray et al. [2012] study used a more sensitive marker of interstitial fibrosis in measuring the average thickness of the alveolar connective tissue, while the DeLorme et al. [2012] study did not report using that assay.

A.7.4. Effects in Other Tissues

In rats, CNF were observed in the nasal turbinates of the high-dose group (25 mg/m^3) at 1 day post-exposure, which was accompanied by hyaline droplet formation in the epithelium; CNF persisted in the nasal turbinates at 3-mo. post-exposure in the high dose group [DeLorme et al. 2012]. In all exposure groups, CNF translocated to the tracheobronchial lymph nodes and CNF fibers were seen in brain, heart, liver, kidney, spleen, intestinal tract, kidneys, and mediastinal lymph nodes, but no associated histopathologic abnormalities were detected [DeLorme et al. 2012]. In CNF-exposed mice, T cell mitogen (concavalin A) responsiveness indicated decrease T cell responses in the spleen at 28 days post-exposure [Murray et al. 2012].

A.7.5. Equivalent Lung Dose Estimation Methods

In order to quantitatively compare the results of the two CNF studies in mice and rats, equivalent lung doses were estimated by accounting for differences in route of exposure and particle size characteristics and by normalizing to either the mass or alveolar surface area of the lungs in each species. The respiratory tract region where the adverse effects were observed is the pulmonary (a.k.a. alveolar) region, which is where gas exchange occurs between the lungs and blood circulatory system across the alveolar septal walls. In mice, the lung dose estimate is simply the proportion of the administered dose (by pharyngeal aspiration) [Murray et al. 2012] that is estimated to deposit in the alveolar region. Mercer et al. [2010] reported that 81% of the aspirated MWCNT by pharyngeal aspiration deposited in the alveolar region of the mouse. If this figure applies to the CNF reported in Murray et al. [2012], then approximately 97 μg of the 120 μg administered dose would be deposited in the alveolar region. In the absence of CNF-specific data, 100% alveolar deposition of the administered dose was also assumed.

In rats, the airborne particle size data are used to estimate the inhalable, deposited, and retained lung doses of CNT, based on the exposure concentrations and particle size characteristics reported [DeLorme et al. 2012]. The multipath particle deposition model (MPPD), version 2.90 [ARA 2009], was used to estimate these lung doses. MPPD version 2.11 was originally used to obtain some particle deposition estimates, but some output indicated errors in estimating the tracheobronchial regional deposited dose, which appeared to lower the alveolar deposition estimates. This issue was apparently resolved in the updated version (2.90).

Particle characteristic input values used in MPPD include the mass median aerodynamic diameter (MMAD), geometric standard deviation (GSD), and density. The following MMAD and GSD values were reported by airborne exposure concentrations: 0.54 mg/m^3 (MMAD 1.9 μm; GSD 3.1); 2.5 mg/m^3 (MMAD 3.2 μm; GSD 2.1); and 25 mg/ m^3 (MMAD 3.3 μm; GSD 2.0). The density assumed for this CNF is 0.08 g/ml. Density was not reported in DeLorme et al. [2012] and was obtained from the manufacturer's data analysis sheet, which indicates it is the same material as that reported in DeLorme et al. [2012].

The default breathing rates and parameters were assumed, and inhalability adjustment was selected. In MPPD 2.90, nonspherical particle shape can be taken into account in the respiratory tract deposition estimates, but some of the required input parameters (GSD of structure diameter and length and correlation) were not reported in DeLorme et al. [2012]. So, the spherical particle assumption (aspect ratio of 1.0) was assumed, which may not be unreasonable given that the fiber interception mechanism may be less for CNF structures of length 5.8 μm than for longer fibers. The default breathing parameters (including 0.21 ml tidal volume and 102 breaths/min) may be reasonable for the female Sprague Dawley rats in the DeLorme et al. [2012] study based on similar body weight (300 g) associated with the default values [Kuempel and Castranova 2011], but may be too low for the male Sprague Dawley rats. The average body weights in control rats (air-only exposed) at the end of 13-wk exposure period and the 90-d post-exposure period, respectively, were: 252 and 329 g (females); 520 and 684 g (males) [DeLorme et al. 2012]. The retained lung burden at the end of the 13-wk exposure was also estimated in MPPD 2.90 using the particle size data for each exposure concentration (using MMAD and GSD values reported above).

The lung dose estimates in rats and mice were normalized by the lung weight or alveolar surface area to estimate the equivalent dose across species. The average lung weights of rats were those reported in DeLorme et al. [2012] 1-d post-exposure in the control rats (1.9 g and 1.3 g in males and females, respectively). The average mouse lung weight was 0.15 g [personal communication, A. Shvedova to E. Kuempel, Aug. 2012]. The average alveolar surface area assumed for the rat lungs was 0.4 m^2 [Stone et al. 1992], and that of mice was 0.055 m^2 [Mercer et al. 2010].

The total deposited CNF dose in the alveolar region was estimated in rats in the DeLorme et al. [2012] study in the following equation:

> Deposited lung dose (mg) = Exposure Concentration (mg/m^3) × Duration (hr/d × d/wk × wk) × Minute Ventilation (L/min) × 0.001 m^3/L × 60 min/hr × Alveolar Deposition Fraction

where the exposure concentrations are 0.54, 2.5, or 25 mg/m^3; the duration of exposure is 6 hr/d, 5 d/wk, 13 wk; the minute ventilation is 0.21 L/min; and the alveolar deposition fractions are reported in Section A.7.5.

A.7.6. Equivalent Lung Dose Estimation Results

The inhalable fraction estimates of CNF in rats were 0.79, 0.73, and 0.72, respectively, in rats at the reported particles sizes for concentrations of 0.54 mg/m^3 (MMAD 1.9 μm; GSD 3.1); 2.5 mg/m^3 (MMAD 3.2 μm; GSD 2.1); and 25 mg/m^3 (MMAD 3.3 μm; GSD 2.0) in DeLorme et al. [2012] (based on MPPD v. 2.90 [ARA 2009] as described in Section A.7.4). The alveolar deposition fraction estimates were 0.0715, 0.0608, and 0.054, respectively, for the 0.54, 2.5, and 25 mg/m^3 exposure concentrations.

The normalized dose estimates in mice and rats (as CNF mass per alveolar surface area or mass of lungs) and associated lung responses are shown in Tables A–15 and A–16. In mice, these lung dose estimates are similar to or higher than the deposited lung dose estimate in the rat at the LOAEL (2.5 mg/ m^3), but less than the deposited lung doses estimated in rats at the highest concentration (25 mg/m^3) (Tables A–15 and A–16). The mouse deposited lung burden estimates are higher than the rat retained lung burden estimates at all doses, assuming spherical-particle model clearance in MPPD 2.90 [ARA 2009]. If CNF is cleared in a similar manner as that reported for MWCNT in Pauluhn [2010b], the actual retained lung dose in rats may be intermediate between the estimated deposited and retained lung burdens. Thus, the mouse fibrotic lung response was observed at an administered lung dose that was similar to, or higher than, the rat lung doses estimated at the LOAEL. This suggests a roughly similar dose-response relationship in the rat and mouse lungs to CNT, based on the limited data in these two studies.

As discussed above (Section A.7.3), the mouse lung responses to CNF (at a 120 µg dose) included alveolar septal thickening identified as pulmonary fibrosis based on collagen deposition observed by Sirius Red staining and the measured thickness of the alveolar connective (septal) tissue [Murray et al. 2012]. In the DeLorme et al. [2012] study, similar qualitative lung responses were observed at the 25 mg/m^3 (as discussed in Section A.7.3). The DeLorme et al. [2012] did not report fibrosis at 25 mg/m^3 although the description of the responses is consistent with early stage fibrosis reported in the Murray et al. [2012].

NOAELs were reported for one type of CNF in De- Lorme et al. [2012] and for one type of MWCNT in Pauluhn [2010a], which were 0.1 and 0.54 mg/m^3, respectively. It follows that the human-equivalent working lifetime exposure estimates to the NO- AEL would be roughly 5-fold higher for the CNF than that for the MWCNT (although not exactly, due to particle size differences and lung deposition estimates). Table A–13 shows estimates of human-equivalent concentrations to effect levels in the Pauluhn and Ma-Hock subchronic inhalation studies, based on different assumptions in extrapolating the rat lung dose to humans. The application of uncertainty factors (e.g., Table A–14) with the CNF used in the DeLorme et al. [2012] study would result in estimated working lifetime no-effect levels in humans of roughly 1–4 $\mu g/m^3$.

Table A–15. CNT lung dose normalized by alveolar surface area in rats and mice

Species and dose*	Deposited lung dose† (mg/m^2 lung)	Retained lung dose† (mg/m^2 lung)	Lung response
Rat: exposure concentration (mg/m^3)			
0.54	0.47	0.084	NOAEL
2.5	1.9	0.25	LOAEL
25	16	1.1	Septal thickening (sligh grade 2) & hypertrophy hyperplasia of type II pneumocytes
Mouse: administered dose (µg)			
120	2.2‡	nd	Septal thickening and pulmonary fibrosis

Abbreviations: NOAEL=no observed adverse effect level; LOAEL= lowest observed adverse effect level; nd=not determined.

* Study references: rat [DeLorme et al. 2012]; mouse [Murray et al. 2012].

† In rats, the pulmonary deposition fraction and 13-wk retained lung burdens were estimated from MPPD 2.9 [ARA 2009].

‡ In mice, this estimate assumes 100% alveolar deposition of the administered by pharyngeal aspiration. If 81% alveolar deposition is assumed as for MWCNT [Mercer et al. 2010], this estimate would be 1.8 mg/m^2 lung.

Table A–16. CNT dose normalized by lung weight in rats and mice

Species and dose*	Deposited dose (mg/g lung)†	Retained dose (mg/g)†
Rat (male): exposure concentration (mg/m^3)		
0.54	0.10	0.018
2.5	0.40	0.054
25	3.4	0.24
Rat (female): exposure concentration (mg/m^3)		
0.54	0.14	0.025
2.5	0.55	0.074
25	4.7	0.33
Mouse (female): administered dose (μg)		
120	0.80‡	nd

Abbreviation: nd=not determined.

* Study references: rat [DeLorme et al. 2012]; mouse [Murray et al. 2012].

† In rats, the pulmonary deposition fraction and 13-wk retained lung burdens were estimated from MPPD 2.9 [ARA 2009].

‡ In mice, this estimate assumes 100% alveolar deposition of the administered by pharyngeal aspiration. If 81% alveolar deposition is assumed as for MWCNT [Mercer et al. 2010], this estimate would be 0.65 mg/g lung.

Appendix B: Occupational Health Surveillance: Informing Decisions Concerning Medical Surveillance in Workplaces with Potential Exposure to CNT and CNF

B.1. Key Terms Related to Medical Surveillance

Occupational health surveillance involves the ongoing systematic collection, analysis, and dissemination of exposure and health data on groups of workers for the purpose of preventing illness and injury. Occupational health surveillance, which includes hazard and medical surveillance, is an essential component of an effective occupational safety and health program [Harber et al. 2003; Baker and Matte 2005; NIOSH 2006; Wagner and Fine 2008; Trout and Schulte 2009], and NIOSH continues to recommend occupational health surveillance as an important part of an effective risk management program.

Hazard surveillance includes elements of hazard and exposure assessment

- The hazard assessment involves reviewing the best available information concerning toxicity of materials. Such an assessment may come from databases, texts, and published literature or available regulations or guidelines (e.g., from NIOSH or the Occupational Safety and Health Administration [OSHA]). Human studies, such as epidemiologic investigations and case series or reports, and animal studies may also provide valuable information. In most instances involving CNT there are limited toxicological data and a lack of epidemiologic data with which to make a complete hazard assessment.

- The exposure assessment involves evaluating relevant exposure routes (inhalation, ingestion, dermal, and/or injection), amount, duration, and frequency (i.e., dose), as well as whether exposure controls are in place and how protective they are. When data are not available, this will be a qualitative process.

B.2. Medical Surveillance

Medical surveillance targets actual health events or a change in a biologic function of an exposed person or persons. Medical surveillance involves the ongoing evaluation of the health status of a group of workers through the collection and aggregate analysis of health data for the purpose of preventing disease and evaluating the effectiveness of intervention programs (primary prevention). NIOSH recommends the medical surveillance of workers when they are exposed to hazardous materials, and therefore are at risk of adverse health effects from such exposures. Medical screening is one form of medical surveillance that is designed to detect early signs of work-related illness in individual workers by administering tests to apparently healthy persons to detect those with early stages of disease or risk of disease. Medical screening generally represents secondary prevention.

Medical surveillance is a second line of defense behind the implementation of engineering, administrative, and work practice controls (including personal protective equipment). Integration of hazard and medical surveillance is important to an effective occupational health surveillance program, and surveillance of disease or illness should not proceed without having a hazard surveillance program in place.

B.2.1. Planning and Conducting Medical Surveillance

Important factors when considering medical surveillance include the following:

1. A clearly defined purpose or objective.
2. A clearly defined target population.
3. The availability of testing modalities to accomplish the defined objective. Testing modalities may include such tools as questionnaires, physical examinations, and medical testing.

A clear plan should be established before beginning a medical surveillance program. The plan should include the following:

1. A rationale for the type of medical surveillance.
2. Provisions for interpreting the results.
3. Presentation of the findings to workers and management of the affected workplace.
4. Implementation of all the other steps of a complete medical surveillance program [Harber et al. 2003].

The elements for conducting a medical surveillance program generally include the following:

1. An initial medical examination and collection of medical and occupational histories.
2. Periodic medical examinations at regularly scheduled intervals, including specific medical screening tests when warranted.
3. More frequent and detailed medical examinations as indicated, based on findings from these examinations.
4. Post-incident examinations and medical screening following uncontrolled or nonroutine increases in exposures such as spills.
5. Worker training to recognize symptoms of exposure to a given hazard.
6. A written report of medical findings.
7. Employer actions in response to identification of potential hazards.

APPENDIX C: NIOSH METHOD 5040

C.1. Background

NIOSH Method 5040 is based on a thermal-optical analysis technique [Birch and Cary 1996] for organic and elemental carbon (OC and EC). The analysis quantifies total carbon (TC) in a sample as the sum of OC and EC. The method was developed to measure diesel particulate matter (DPM) in occupational settings, but it can be applied to other types of carbonaceous aerosols. It is widely used for environmental and occupational monitoring.

For the thermal-optical analysis, a portion (typically a 1.5-cm^2 rectangular punch) of a quartz-fiber filter sample is removed and placed on a small quartz spatula. The spatula is inserted in the instrument's sample oven, and the oven is tightly sealed. Quartz-fiber filters are required for sample collection because temperatures of 850 °C and higher are employed during the analysis. The thermal-optical analyzer is equipped with a pulsed diode laser and photo detector that permit continuous monitoring of the filter transmittance. This optical feature corrects for the "char" that forms during the analysis because of carbonization of some materials.

Thermal-optical analysis proceeds in inert and oxidizing atmospheres. In both, the evolved carbon is catalytically oxidized to carbon dioxide (CO_2). The CO_2 is then reduced to methane (CH_4), and CH_4 is quantified with a flame ionization detector (FID). The OC (and carbonate, if present) is first removed in helium, as the temperature is increased to a preset maximum. If sample charring occurs, the filter transmittance decreases as the temperature is stepped to the maximum. After the OC is removed in helium, an oxygen-helium mix is introduced, and the temperature is again stepped to a maximum (850 °C or higher, depending on the sample) to effect combustion of the remaining material. As the light-absorbing carbon (mainly EC and char) is oxidized from the filter, the filter transmittance increases. The split between the OC and EC is assigned when the initial (baseline) value of the filter transmittance is reached. All carbon removed before the OC-EC split is considered organic, and that removed after the split is considered elemental.

If no charring occurs, the split is assigned before removal of EC. If the sample chars, the split is not assigned until enough light-absorbing carbon is removed to increase the transmittance to its initial value.

OC and EC results are reported as micrograms per square centimeter (μg/cm^2) of sample deposit. The total OC and EC on the filter are calculated by multiplying the reported values by the deposit area. Because only a portion of the sample is analyzed, it must be representative of the entire deposit. Thus, a homogeneous deposit is assumed. The entire filter must be analyzed (in portions if a 37-mm filter is used) if the filter deposit is uneven.

C.2. Method Evaluation

The reported accuracy of NIOSH 5040 is based on analysis of TC in different sample types. Accuracy was based on TC, because there is no analytical standard for determining the OC-EC content of a complex carbonaceous aerosol. In the method evaluation, five different organic compounds were analyzed to examine whether the instrument response is compound dependent. Linear regression of the data (43 analyses total) for all five compounds gave a slope and correlation coefficient (r) near unity [slope = 0.99 (± 0.01), r^2 = 0.999, n = 43], indicating a compound-independent response. Eight different carbonaceous materials also were analyzed by three methods, in-house by thermal-optical analysis and by two other methods used by two external laboratories. Sample materials included, DPM, coals, urban dust, and humic acid. Thermal-optical results agreed well with those reported by the two other laboratories. The variability of the TC results for the three laboratories ranged from about 1%–7%. These findings [Birch and Cary 1996] demonstrate that carbon can be accurately quantified irrespective of the compound or sample type.

In sampling DPM, different samplers gave comparable EC results because particles from combustion sources are generally less than one μm (diameter). As such, the particles are collected with high efficiency (near 100%) and evenly deposited on the filter. In the method evaluation, different sampler types (open-face 25-mm and 37-mm cassettes, 298 personal cascade impactors, and four proto- type impactors) were used to collect diesel exhaust aerosol at an express mail facility. The relative standard deviation (RSD) for the mean EC concentration was 5.6% [Birch and Cary 1996]. Based on the 95% confidence limit (19%; 13 degrees of freedom, n = 14) on the accuracy, the NIOSH accuracy criterion [Kennedy et al. 1995] was fulfilled. Variability for the OC results was higher (RSD = 12.3%), which is to be expected when different samplers are used to collect aerosols that contain semi-volatile (and volatile) components, because these may have a filter face velocity dependence. The method pre- cision (RSD) for triplicate analyses (1.5 cm^2 filter portions) of a 37-mm quartz-fiber filter sample of DPM was normally better than 5%, and often 2% or less [NIOSH 1994a].

In the method evaluation, the limit of detection (LOD) was estimated two ways: (1) through analysis of low-level calibration standards [Birch and Cary 1996; NIOSH 1994], and (2) through analysis of pre-cleaned media blanks. In the first approach, OC standard solutions (sucrose and ethylene diaminetetraacetic acid [EDTA]) covering a range from 0.23 to 2.82 μg C (or from 0.15 to 1.83 μg C per cm^2 of filter) were analyzed. An aliquot (usually 10 μL) of the standard was applied to one end of a 1.5-cm^2 rectangular filter portion that was pre-cleaned in the sample oven just before application of the aliquot. The filter portion was pre-cleaned to remove any OC contamination, which can greatly increase the EC LOD when TC results are used for its estimation. After cleaning the filter portion, metal tweezers are used to remove from the sample oven the quartz spatula that holds the portion. External to the oven, the spatula is held in place by a metal bracket such that the standard can be applied without removing the filter portion from the spatula. This avoids potential contamination from handling.

Results of linear regression of the low-level calibration data were used to calculate the LOD as 3 σy/m, where σy is the standard error of the regression and m is the slope of the regression line. TC results were used rather than OC because the pyrolysis correction may not account for all of the char formed during analysis of the standard (because of low sample loading and/or the position of the aliquot in the laser). If not, a small amount of the OC will be reported as EC, introducing variability in the OC results and increasing the LOD. The LOD estimated through the linear regression results was 0.24 μg C per filter portion, or 0.15 μg/cm^2.

A simpler approach for LOD determination is through analysis of media blanks. In the method evaluation, TC results for pre-cleaned, 1.5-cm^2 portions of the filter media were used to calculate the LOD estimate. The mean (n = 40) TC blank was 0.03 ±0.1 μg TC. Thus, the LOD estimated as three times the standard deviation for pre-cleaned media blanks (3 σ blank) was about 0.3 μg C. This result agrees well with the value (0.24 μg C) estimated through analysis of the standard solutions. Considering a 960-L air sample collected on a 37-mm filter and a 1.5-cm^2 sample portion, this LOD translates to an air concentration of about 2 μg/m^3 ([0.3 μg TC/1.5 cm^2] [8.5 cm^2]/0.960 m^3 = 1.78 μg/m^3), corresponding to the reported upper LOQ of about 7 μg/m^3 (LOQ = 3.3 × LOD).

As with all analytical methods, the LOD is a varying number. However, the EC LOD (about 2 μg/m^3, or an LOQ of 7 μg/m^3) reported for NIOSH Method 5040 is a high estimate. As discussed in Section 6 of the CIB, it was based on analysis of pre-cleaned media blanks from different filter lots, over a 6-month period, and by different analysts at two different laboratories. Further, variability for the TC results, rather than the EC results, was used to estimate the LOD. These combined factors gave a conservative (high) estimate of the EC LOD. More typical values, under different sampling conditions, are discussed in Section 6.1 of the CIB.

C.3. Inter-Laboratory Comparisons

When results of the initial method evaluation were published [Birch and Cary 1996], an interlaboratory comparison was not possible because the thermal-optical instrument was available in only one laboratory. After additional laboratories acquired thermal-optical instruments, a round robin comparison [Birch 1998] was conducted. Matched sets of filter samples containing different types of complex carbonaceous aerosols were distributed to 11 laboratories. Six of the eleven analyzed the samples according to NIOSH 5040, while five used purely thermal (i.e., no char correction) methods. Good interlaboratory agreement was obtained among the six laboratories that used NIOSH 5040. In the analysis of samples containing DPM, the variability (RSD) for the EC results ranged from 6% to 9%. Only low EC fractions were found in wood and cigarette smoke. Thus, these materials pose minimal interference in the analysis of EC. In addition, only minor amounts of EC were found in two OC standards that char: about 1% for sucrose and 0.1% for the disodium salt of ethylene diaminetetraacetic acid (EDTA). Two aqueous solutions of OC standards were included in the comparison as a check on the validity of the char correction and accuracy of the TC results.

Variability (RSD) of the TC results for the two standard solutions and five filter samples ranged from 3% to 6%.

A second interlaboratory comparison study using NIOSH 5040 was also conducted [Schauer et al. 2003]. Seven environmental aerosol samples were analyzed in duplicate by eight laboratories. Four samples were collected in U.S. cities, and three were collected in Asia. Interlaboratory variability for the EC results ranged from 6% to 21% for six samples having EC loadings from 0.7 to 8.4 $\mu g/cm^2$. Four of the six had low EC loadings (0.7 $\mu g/cm^2$ to 1.4 $\mu g/cm^2$). The variability for the OC results ranged from 4% to 13% (OC loadings ranged from about 1 to 25 $\mu g/cm^2$). Results for TC were not reported, but the variability reported for the OC results should be representative of that for TC, because the samples were mostly OC (75% to 92%). Similar findings were also reported by Chai et al. [2012] from seven laboratories in which analysis was performed using Method 5040 on four sample filter sets containing OC and EC. The summary RSDs for EC results were <12% for all four sample sets.

C.4. Carbonates

Carbonate in a sample is indicated by a relatively narrow peak during the fourth temperature step in helium [Birch 2004a]. Its presence is verified by exposing a second portion of the filter to hydrogen chloride (HCL) vapor before analysis. When the acidified portion is analyzed, a diminished (or absent) peak during the fourth temperature step is indicative of carbonate in the original sample [Birch 2004a]. Environmental samples typically contain little (if any) carbonate, but carbonate (e.g., in limestone, trona, concrete) levels in some occupational samples can be quite high. In such cases, it is important to ensure that all of the carbonate is removed during the first stage of the analysis. If it is not completely removed (because of high loading), the sample should be acidified.

C.5. Organic Carbon Sampling Artifacts

Problems commonly referred to as "sampling artifacts" have been reported when collecting particulate OC on quartz fiber filters. These artifacts do not affect the EC results, but they cause positive or negative bias in the measurement of particulate OC (and TC). Eatough et al. [1995, 1996] observed loss of semi-volatile OC from particles during sampling, referred to as the "negative" or evaporation artifact. This artifact causes a negative bias in the particulate OC (and TC) concentration, because OC initially collected as condensed matter is subsequently lost through evaporation from the filter during sampling. Conversely, several studies have demonstrated a "positive" or adsorption artifact because of filter adsorption of gas phase OC. A quartz-fiber filter collects airborne particulate matter and allows gases and vapors to pass through, but some adsorption of gas phase (and vapor) OC occurs, resulting in overestimation of the true airborne particulate OC concentration [Turpin et al. 2000; McDow and Huntzicker 1990; Turpin et al. 1994; Olson and Norris 2005; Kirchstetter et al. 2001; Mader et al. 2003; Subramanian et al. 2004; Mader et al. 2001; Noll and Birch 2008; Schauer et al. 1999].

Most of the studies on sampling artifacts apply to environmental air sampling. Occupational sampling methods and conditions are generally much different than environmental. Environmental samples are usually collected at much higher face velocities: 20–80 cm/s as opposed to 3–4 cm/s for occupational samples. In addition, the concentrations of carbon are much lower in environmental air than in most occupational settings [Fruin et al. 2004; Sheesley et al. 2008], and the types of aerosols sampled are different (e.g., aged aerosol from multiple environmental sources, as opposed to aerosols close to source). These differences are important because OC sampling artifacts depend upon conditions such as filter face velocity, air contaminants present, sampling time, and filter media. Given the much lower filter face velocities typical of occupational sampling, adsorption (i.e., positive artifact) is expected over evaporation for occupational samples. Turpin et al. [1994], Kirchstetter et al. [2001], Noll and Birch [2008], and Schauer et al. [1999] have reported adsorption as the dominant artifact.

To correct for the positive adsorption artifact, tandem quartz filters have been applied. When sampling with tandem filters, particulate matter is collected by the first filter, while both the first and second filters are exposed to and adsorb gaseous and vaporous OC. For the correction to be effective, both filters must be in equilibrium with the sampled airstream, adsorb the same amount of gas/ vapor OC, and not have a significant amount of OC loss through evaporation. The OC on the second filter can then be subtracted from the OC on the first filter to account for the adsorbed OC. Several studies have found the tandem filter correction to underestimate the adsorption artifact [Turpin et al. 2000; McDow and Huntzicker 1990; Turpin et al. 1994; Olson and Norris 2005], while others have shown effective correction [Kirchstetter et al. 2001; Mader et al. 2003; Subramanian et al. 2004; Mader et al. 2001; Noll and Birch 2008].

Air samplers containing a Teflon® and quartz filter also have been used for correction of the positive OC artifact. In theory, the Teflon top filter collects particulate matter with negligible OC gas/vapor adsorption, so only the quartz filter beneath it adsorbs gas and vapor OC. Studies on tandem filter corrections have shown the quartz filter beneath Teflon to have a greater OC value than quartz be- neath quartz [Turpin et al. 2000; Olson and Norris 2005]. This finding was attributed to the quartz beneath quartz not reaching equilibrium with the sampling stream and underestimating the adsorption artifact. Others have attributed it to the evaporation artifact being more prevalent when using a Teflon filter instead of a quartz filter, and they reported the quartz behind Teflon to overestimate the adsorption artifact [Subramanian et al. 2004]. Several studies have shown no difference when using either type of correction [Mader et al. 2003; Mader et al. 2001].

Noll and Birch [2008] conducted studies on OC sampling artifacts for occupational samples to test the accuracy of the tandem quartz-filter correction. In practice, using two quartz filters for air sampling is preferable to the Teflon-quartz combination because both the collection and blank filters are in the same sampler. The tandem quartz correction effectively reduced positive bias for both laboratory and field samples. Laboratory samples were collected under conditions that simulated DPM sampling in underground mines. Without correction, TC on the sample filter was 30% higher than the actual particulate TC for 50% of the samples, but it was within 11% of the particulate TC after the tandem quartz-fiber correction. For field samples, this correction significantly reduced positive bias due to OC adsorption artifact. Little artifact effect was found after the correction was made.

C.6. Carbon Nanotubes and Nanofibers

Method 5040 was developed to measure DPM in occupational environments, but it can be applied to other types of carbonaceous aerosols. When applied to materials such as carbon black or CNT/ CNF, particle deposition on a filter may be more variable because particles in these materials are much larger than DPM. Variability depends on the sampler type, and as expected, different samplers (e.g., cyclones, open- and closed-face cassettes) will give different air concentration results, depending on the particle size distribution [Birch et al. 2011]. Diesel emissions, and combustion aerosols generally, are composed of ultrafine (< 100 nm diameters) particles. Because of the small size, DPM normally deposits evenly across the quartz-fiber filter used for sample collection. As already discussed, even deposition is required because only a portion of the filter is normally analyzed. Thus, it must be representative of the entire sample deposit.

When applying NIOSH 5040 to CNT/CNF, it is important to verify an even filter deposit so that an accurate air concentration (based on results for the filter portion) can be calculated. Alternatively, the entire filter can be analyzed if the deposit is uneven, but this requires analysis of multiple portions of a 37-mm filter because of the relatively small diameter (about 1 cm) of the carbon analyzer's quartz sample oven. Quality assurance procedures should include duplicate analyses of the 37-mm filter to check precision, especially if the deposit appears uneven. If a 25-mm filter is used, the entire filter can be analyzed, which improves the LOD and obviates the need for an even deposit, but a repeat analysis (or other chemical analysis) of the sample is not possible if the entire filter is analyzed. In addition, the filter must be cut into portions, and the portions must be properly loaded in the analyzer so the sample transmittance can be monitored. Additional details on the evaluation and use of NIOSH 5040 are provided elsewhere [Birch and Cary 1996; Birch 1998; Birch et al. 1999; Birch 2002; Birch 2003; Birch 2004a].

As discussed in the CIB, NIOSH 5040 has been applied to several field studies on CNT/CNF [Methner et al. 2007; Birch et al. 2011b; Dahm et al. 2011]. In one study, it was employed for area monitoring at a laboratory facility that processes CNF in the production of polymer composites [Methner et al. 2007]. Carbon nanofibers and CNT have negligible (if any) OC content, making EC a good indicator of these materials. Survey results were reported in terms of TC, which is subject to OC interferences, but the OC results were blank corrected by the tandem filter method described in the preceding section (organic carbon adsorption artifacts) to minimize the positive sampling artifact. Further, based on the thermal profiles for the air samples and the bulk materials (CNF and composite product), the blank-corrected TC was a good measure of the CNF air concentration, except in an area where a wet saw was operating. In that area, TC was a measure of the composite aerosol released during the sawing operation, which contained a high OC fraction due to the composite matrix.

There are several issues and limitations when analyzing dusts generated during cutting, sanding, or grinding CNT/CNF composites. First, the accuracy of determining the EC fraction of a polymer composite is questionable and expected to vary, depending on polymer type and sample loading. Further, EC in both the polymer and bulk CNT/CNF materials will be measured (i.e., not speciated) if both are present. In addition, the EC loading in a polymer composite is usually a low percentage (e.g., 1%). Therefore, if the composite dust is the only EC source, and if its EC content is determined accurately, an EC concentration of 2 $\mu g/m^3$ would correspond to a dust concentration (at 1% EC) of 200 $\mu g/m^3$, considerably higher than the EC concentration. As such, the sample can be easily overloaded with OC because of the high relative OC content, which can both overload the analyzer and cause positive bias in the EC result. Further, in a composite particle, the CNT/CNF is bound within a polymer (or resin) matrix, which is dissimilar to a particle of unbound material. An effort to improve the analysis of samples containing dusts of polymer composites is ongoing; however, in the context of the NIOSH REL, the intended application is CNT/CNF in powder form, purified or unpurified. Whenever possible, a bulk sample of the material (and, if available, other materials that may be aerosolized) should be analyzed as the thermal properties of CNT/CNF are material dependent (e.g., CNF, SWCNT, MWCNT, functionalized or not functionalized). The OC-EC split for a bulk material is not reliable because it depends on how the powder is applied to the filter punch, but a small amount of the CNT/CNF should be analyzed to determine the onset of oxidation of the material and confirm its complete oxidation.

Table 1. NIOSH 5040 precision for air samples collected in different areas of a CNF manufacturing facility with 37-mm cassettes (total dust) and cyclone samplers (thoracic and respirable dust). OC, EC, TC are reported as air concentrations (μg C/m^3)

Sample	OC*	RPD or RSD† (%)	EC‡	RPD or RSD (%)	TC	RPD or RSD (%)	Comments
Respirable	16.42	0.97	[1.87]§	13.37	18.28	2.19	paired¶
Respirable	22.19	8.25	3.41	22.29	25.66	10.60	paired
Total	27.17	13.40	21.52	12.04	48.69	12.80	duplicate**
Respirable	60.87	0.74	79.59	12.14	140.31	6.36	duplicate
Respirable	25.47	4.46	20.72	8.48	46.09	6.28	duplicate
Total	12.42	6.84	4.14	4.59	16.60	4.88	triplicate††
Respirable	19.89	3.22	3.05	4.59	22.93	2.22	triplicate
Total	15.11	1.29	9.89	9.37	25.01	3.63	triplicate
Total	17.80	9.72	11.07	7.97	28.88	9.15	paired
Thoracic	27.16	10.80	11.23	6.79	38.46	6.68	triplicate
Respirable	22.81	2.50	23.67	13.86	46.48	8.26	duplicate
Respirable	18.64	6.77	8.44	3.15	27.14	5.63	duplicate

*OC = organic carbon.
† RPD is relative percent difference. RSD is relative standard deviation.
‡ EC = elemental carbon.
§ Result in brackets lies between method LOD and LOQ.
¶ Results for two identical, paired samplers.
** Duplicate analysis of same filter.
†† Triplicate analysis of same filter.

NIOSH investigators also conducted extensive air monitoring at a facility that manufactures and processes CNFs [Evans et al. 2010; Birch 2011a; Birch et al. 2011b]. Both personal breathing zone and area samples were collected. To evaluate the method precision, paired samples were collected and repeat analyses of the filters were performed. The relative percent difference (RPD) and RSD (%) for repeat analyses of 12 samples collected in different areas of the facility are listed in Table 1. Total, thoracic, and respirable dust samples are included. Total (inhalable equivalent) dust was collected with 37-mm cassettes, while cyclones were used to collect thoracic and respirable dust. The RPD was determined by analyzing either two punches from the same filter (duplicates) or one punch from two different filters (paired samplers); the RSD was determined by analyzing one filter in triplicate. The precision for the EC results ranged from about 3% to 14% except for one respirable sample, where the RPD was about 22%. Higher variability for the latter may relate to spatial variation, because the two filter punches analyzed were from different samplers. Spatial variation, rather than sampler variability, is a likely explanation for this particular result as two other sets of paired samplers do not show higher variability. The RPDs for these are about 8% and 13%, comparable to results for multiple punches from the same filter.

REFERENCES

ACGIH. (1984). Particle size-selective sampling in the workplace. Report of the ACGIH Technical Committee on air sampling procedures. *Ann Am Conf Gov Ind Hyg*, *11*, 23–100.

ACGIH. (2007). Industrial ventilation: a manual of recommended practice. 26th ed. Cincinnati, OH.

AISE. (2001). Guidelines for the safe handling of enzymes in detergent manufacturing [http://www. aise.eu/downloads/AISE_Guidelines_for_the_ safe_handling_of_enzymes_ in_detergent_manufacturing.pdf].

Aiso, S., Yamazaki, K., Umeda, Y., Asakura, M., Kasai, T., Takaya, M., Toya, T., Koda, S., Nagano, K., Arito, H. & Fukushima, S. (2010). Pulmonary toxicity of intratracheal instilled multiwall carbon nanotubes in male Fischer 344 rats. *Ind Health*, *48*(6), 783–795.

Allen, B. L., Kichambare, P. D., Gou, P., Vlasova, I. I., Kapralov, A. A., Konduru, N., Kagan, V. E. & Star, A. (2008). Biodegradation of single-walled carbon nanotubes through enzymatic catalysis. *Nano Lett*, *8*(11), 3899–3903.

Anjilvel, S. & Asgharian, B. (1995). A multiple-path model of particle deposition in the rat lung. *Fundam Appl Toxicol*, *28*(1), 41–50.

ARA. (2011). Multiple-path particle deposition (MPPD 2.1): a model for human and rat airway particle dosimetry. Raleigh, NC: Applied Research Associates, Inc.

Arkema Inc. (2008). Letter of June 2, 2008, from Debra Randall, Manager, Product Safety, Arkema Inc., to TSCA Document Control Officer, U.S. Environmental Protection Agency [http:// www.epa.gov/oppt/tsca8e/pubs/8ehq/2008/ jun08/8ehq_0608_17109b.pdf].

Asakura, M., Sasaki, T., Sugiyama, T., Takaya, M., Koda, S., Nagano, K., Arito, H. & Fukushima, S. (2010). Genotoxicity and cytotoxicity of multi-wall carbon nanotubes in cultured Chinese hamster lung cells in comparison with chrysotile A fibers. *J Occup Health*, *52*(3), 155–166.

Aschberger, K., Johnston, H. J., Stone, V., Aitken, R. J., Hankin, S. M., Peters, S. A., Tran, C. L. & Christensen, F. M. (2010). Review of carbon nanotubes toxicity and exposure—appraisal of human health risk assessment based on open literature. *Crit Rev Toxicol*, *40*(9), 759–790.

Asgharian, B., Wong, B. A., Bermundez, E. & Everitt, J. I. (2001). A model of clearance of titanium dioxide from the rat lung. *The Toxicologist* (supplement to Toxicol Sci), *60*(1), 195.

Asgharian, B., Kelly, J. T. & Tewksbury, E. W. (2003). Respiratory deposition and inhalability of mono-disperse aerosols in Long-Evans rats. *Toxicol Sci*, *71*(1), 104–111.

Bailey, M., Ansoborlo, E., Etherington, G., Gregoratto, D., Guilmette, R., Marsh, J., Paquet, F. & Smith, J. (2008). Proposed updating of the ICRP human respiratory tract model. Presented at the 12th International Congress of the International Radiation Protection Association [http://www.irpa12.org.ar/fullpapers/ FP0947.pdf].

Baker, E. L. & Matte, T. P. (2005). Occupational health surveillance. In: Rosenstock L, Cullen E, Brodkin R, eds. Textbook of clinical occupational and environmental medicine. Philadelphia, PA: Elsevier Saunders Company [http://www.osha.gov/SLTC/ medicalsurveillance/surveillance.html].

Bałazy, A., Toivola, M., Reponen, T., Podgorski, A., Zimmer, A. & Grinshpun, S. A. (2006). Manikin-based performance evaluation of N95 filtering facepiece respirators challenged with nanoparticles. *Ann Occup Hyg*, *50*(3), 259–269.

Barillet, S., Simon-Deckers, A., Herlin-Boime, N., Mayne, L., L'Hermite, M., Reynaud, C., Cassio, D., Gouget, B. & Carriere, M. (2010). Toxicological consequences of TiO_2, SiC nanoparticles and multi-walled carbon nanotubes exposure in several mammalian cell types: an in vitro study. *J Nanopart Res*, *12*(1), 61–73.

Baron, P. A., Deye, G. J., Chen, B. T., Schwegler-Berry, D., Shvedova, A. A. & Castranova, V. (2008). Aerosolization of single-walled carbon nanotubes for an inhalation study. *Inhal Toxicol*, *20*(8), 751–760.

Bello, D., Hart, J. A., Ahn, K., Hallock, M., Yamamoto, N., Garcia, E. J., Ellenbecker, M. J. & Wardle, B. L. (2008). Particle exposure levels during CVD growth and subsequent handling of vertically-aligned carbon nanotube films. *Carbon*, *46*(6), 974–981.

Bello, D., Wardle, B. L., Yamamoto, N., de Villora, R. G., Garcia, E. J., Hart, A. J., Ahn, K., Ellenbecker, M. J. & Hallock, M. (2009). Exposure to nanoscale particles and fibers during machining of hybrid advanced composites containing carbon nanotubes. *J Nanopart Res*, *11*(1), 231–249.

Bello, D., Wardle, B. L., Zhang, J., Yamamoto, N., Santeufemio, C., Hallock, M. & Virji, M. A. (2010). Characterization of exposures to nanoscale particles and fibers during solid core drilling of hybrid carbon nanotubes advanced composites. *Int J Occup Environ Health*, *16*(6), 434–450.

Bellmann, B., Muhle, H., Creutzenberg, O. & Mermelstein, R. (1990). Recovery behavior after dust overloading of lungs in rats. *J Aerosol Sci*, *21*(3), 377–380.

Bellmann, B., Muhle, H., Creutzenberg, O., Dasen-brook, C., Kilpper, R., MacKenzie, J. C., Morrow, P. & Mermelstein, R. (1991). Lung clearance and retention of toner, utilizing a tracer technique, during chronic inhalation exposure in rats. *Fundan Appl Toxicol*, *17*, 300–313.

Birch, M. E. & Cary, R. A. (1996). Elemental carbon-based method for monitoring occupational exposure to particulate diesel exhaust. *Aeros Sci and Technol*, *25(*3), 221–241.

Birch, M. E. (1998). Analysis of carbonaceous aerosols: interlaboratory comparison. *Analyst*, *123*(5), 851–857.

Birch, M. E., Dahmann, D. & Fricke, H.-H. (1999). Comparison of two carbon analysis methods for monitoring diesel particulate levels in mines. *J Environ Monit*, *1*(6), 541.

Birch, M. E. (2002). Occupational monitoring of particulate diesel exhaust by NIOSH Method 5040. *Appl Occup Environ Hyg*, *17*(6), 400–405.

Birch, M. E. (2003). Standard test method for monitoring diesel particulate exhaust in the workplace. ASTM standard test method D 6877-03. West Conshohocken, PA: ASTM International.

Birch, M. E. (2004a). Monitoring of diesel particulate exhaust in the workplace. Chapter Q. In: Schlecht PC, O'Connor PF, eds. NIOSH manual of analytical methods (NMAM). 4th ed. Cincinnati, OH: U.S. Department of Health and Human Services, Centers for Disease Control and Prevention, National Institute for Occupational Safety and Health, DHHS (NIOSH) Publication No. 2003–154.

Birch, M. E. (2004b). NIOSH Method 5040 update. In: Schlecht PC, O'Connor PF, eds. NIOSH manual of analytical methods (NMAM). 3rd supplement, 4th Edition. Cincinnati, OH: U.S. Department of Health and Human Services, Centers for Disease Control and Prevention, National Institute for Occupational Safety and Health, DHHS (NIOSH) Publication No. 2003–154.

Birch, M. E. (2011a). Exposure and emissions monitoring during carbon nanofiber production–part II: polycyclic aromatic hydrocarbons. *Ann Occup Hyg*, *55*(9), 1035–10470.

Birch, M. E., Ku, B.-K., Evans, D. E. & Ruda-Eberenz, T. (2011b). Exposure and emissions monitoring during carbon nanofiber production–part I: elemental carbon and iron-soot aerosols. *Ann Occup Hyg*, *55*(9), 1016–1036.

Bohning, D. E. & Lippmann, M. (1992). Particle deposition and pulmonary defense mechanisms. Chapter 14, Organ Systems: Lung. *In: Rom WN, ed. Environmental and Occupational Medicine*, 2nd ed., 171–182. Boston, Little, Brown, and Company.

Bolton, R. E., Vincent, H. J., Jones, A. D., Addison, J. & Beckett, S. T. (1983). An overload hypothesis for pulmonary clearance of UICC amosite fibers inhaled by rats. *Br J In Med*, *40*, 264–272.

Brouwer, D., van Duuren-Stuurman, B., Berges, M., Jankowska, E., Bard, D. & Mark, D. (2009). From workplace air measurement results toward estimates of exposure? Development of a strategy to assess exposure to manufactured nanoobjects. *J Nanopart Res*, *11*(8), 1867–1881.

BSI. (2007). Nanotechnologies—Part 2: *Guide to safe handling and disposal of manufactured nanomaterials PD*, 6699-2, 2008. London: BSI Group.

Bullock, W., Ignacio, J. S. eds. (2006). *A strategy for assessing and managing occupational exposures*. 3rd ed. Fairfax, VA: AIHA Press.

Carrero-Sanchez, J. C., Elias, A. L., Mancilla, R., Arrellin, G., Terrones, H., Laclette, J. P. & Terrones, M. (2006). Biocompatibility and toxicological studies of carbon nanotubes doped with nitrogen. *Nano Lett*, *6*(8), 1609–1616.

CEN. (1993). Workplace atmospheres—size fraction definitions for measurement of airborne particles, EN 481. Brussels, Belgium: European Committee for Standardization.

Cena, L. G. & Peters, T. M. (2011). Characterization and control of airborne particles emitted during production of epoxy/carbon nanotube nanocomposites. *J Occup Environ Hyg*, *2*(8), 86–92.

Chai, M., Birch, M. E. & Deye, G. (2012). Organic and elemental carbon filter sets: preparation method and interlaboratory results. *Ann Occup Hyg*, doi:10.1093/annhyg/mes029.

Chang, Ly., Huang, Y., Stockstill, B. L., Graham, J. A., Grose, E. C., Menache, M. G., Miller, F. J., Costa, D. L. & Crapo, J. D. (1992). Epithelial injury and interstitial fibrosis in the proximal alveolar regions of rats chronically exposed to a simulated pattern of urban ambient ozone. *Toxicol Appl Pharmacol*, *115*(2), 241–252.

CIIT & RIVM. (2006). Multiple-path particle dosimetry (MPPD, version 2.0): a model for human and rat airway particle dosimetry. Research Triangle Park, NC: Centers for Health Research (CIIT) and the Netherlands: National Institute for Public Health and the Environment (RIVM).

Coppeta, L., Legramante, J., Galante, A., Bergamashi, A. J., Bergamashi, E., Margrini, A. & Pietroiusti, A. (2007). Interaction between carbon nanotubes and cardio- vascular autonomic nervous system regulation: proposal of an animal model and preliminary findings. *G Ital Med Lav Ergon*, *5*(29), 465–467.

Corn, M. & Esmen, N. (1979).Workplace exposure zones for classification of employee exposure to physical and chemical agents. *Am Ind Hyg Assoc*, *1*(40), 47–57.

Costa, D. L., Tepper, J. S., Stevens, M. A., Watkinson, W. P., Doerfier, D. L., Gelzleichter, T. R. & Last, J. A. (1995). Restrictive lung disease in rats exposed chronically to an urban profile of ozone. *Am J Respir Crit Care Med*, *151*(5), 1512–1518.

Crump, K. S. (1984). A new method for determining allowable daily intakes. *Fund Appl Toxicol*, *4*(5), 854–871.

Crump, K. S. (1995). Calculation of benchmark doses from continuous data. *Risk Anal*, *15*(1), 79–89.

Crump, K. S. (2002). Critical issues in benchmark calculations from continuous data. *Crit Rev Toxicol*, *32*(3), 133–153.

Dahm, M. M., Evans, D. E., Schubauer-Berigan, M. K., Birch, M. E. & Fernback, J. E. (2011). Occupational exposure assessment in carbon nanotube and nanofiber primary and secondary manufactures. *Ann Occup Hyg*, doi:10.1093/annhyg/mer110l

DeLorme, M. P., Muro, Y., Arai, T., Banas, D. A., Frame, S. R., Reed, K. L. & Warheit, D. B. (2012). Ninety-day inhalation toxicity study with a vapor grown carbon nanofiber in rats. *Tox Sci*, *128*(2), 449–460.

Deng, X., Jia, G., Wang, H., Sun, H., Wang, X., Yang, S., Wang, T. & Liu, Y. (2007). Translocation and fate of multi-walled carbon nanotubes in vivo. *Carbon*, *45*(7), 1419−1424.

Donaldson, K., Borm, P. J., Oberdörster, G., Pinkerton, K. E., Stone, V. & Tran, C. L. (2008). Concordance between in vitro and in vivo dosimetry in the proinflammatory effects of low-toxicity, low-solubility particles: the key role of the proximal alveolar region. *Inhal Toxicol*, *20*(1), 53–62.

Driscoll, K. E., Carter, J. M., Howard, B. W., Hassenbein, D. G., Pepelko, W., Baggs, R. B. & Oberdörster, G. (1996). Pulmonary inflammatory chemokine and mutagenic responses in rats after subchronic inhalation of carbon black. *Toxicol Appl Pharmacol*, *136*(2), 372–380.

Duan, N. & Mage, T. (1997). Combination of direct and indirect approaches for exposure assessment. *J Expo Anal Environ Epidemiol*, *7*(4), 439–470.

Eatough, D. J., Tang, H., Cui, W. & Machir, J. (1995). Determination of the size distribution and chemical composition of fine particulate semivolatile organic material in urban environments using diffusion denuder technology. *Inhal Toxicol*, *7*(5), 691–710.

Eatough, D. J., Eatough, D. A., Lewis, L. & Lewis, E. (1996). Fine particulate chemical composition and light extinction at Canyonlands National Park using organic particulate material concentrations obtained with multisystem, multichannel diffusion denuder sampler. *J Geophys Res*, *101*(D14), 19515–19531.

Elder, A., Gelein, R., Finkelstein, J. N., Driscoll, K. E., Harkema, J. & Oberdörster, G. (2005). Effects of subchronically inhaled carbon black in three species. I. Retention kinetics, lung inflammation, and histopathology. *Toxicol Sci*, *88*(2), 614–629.

Elgrabli, D., Abella-Gallart, S., Robidel, F., Rogerieux, F., Boczkowski, J. & Lacroix, G. (2008a). Induction of apoptosis and absence of inflammation in rat lung after intratracheal instillation of multiwalled carbon nanotubes. *Toxicology*, *253*(1–3), 131–136 [Epub ahead of print].

Elgrabli, D., Floriani, M., Abella-Gallart, S., Meuneir, L., Gamez, C., Delalain, P., Rogerieux, F., Boczkowski, J. & Lacroix, G. (2008b). Biodistribution and clearance of instilled carbon nanotubes in rat lung. *Part Fibre Toxicol*, Dec 9, *5*, 20 [http://dx.doi.org/10.1186/1743-8977-8-1510.1186/1743-8977-5-20].

Ellinger-Ziegelbauer, H. & Pauluhn, J. (2009). Pulmonary toxicity of multi-walled carbon nanotubes (Baytubes) relative to quartz following a single 6h inhalation exposure of rats and a 3 months post-exposure period. *Toxicology*, *266*(1–3), 16–29.

Erdely, A., Hulderman, T., Salmen, R., Liston, A., Zeidler-Erdely, P. C., Schwegler-Berry, D., Castranova, V., Koyama, S., Kim, Y. A., Endo, M. & Simeonova, P. P. (2009). Cross-talk between lung and systemic circulation during carbon nanotube respiratory exposure. Potential biomarkers. *Nano Lett*, *1*(9), 36–43.

Evans, D. E., Ku, B.-K., Birch, M. E. & Dunn, K. H. (2010). Aerosol monitoring during carbon nanofiber production: mobile direct-reading sampling. *Ann Occup Hyg*, *54*(5), 514–531.

Fenoglio, I., Greco, G., Tomatis, M., Muller, J., Raymundo- Piñero, Béguin, F., Fonseca, A., Nagy, J. B., Lison, D. & Fubini, B. (2008). Structural defects play a major role in the acute lung toxicity of multiwall carbon nanotubes: physicochemical aspects. *Chem Res Toxicol*, *21*(9), 1690–1697.

Ferris, B. G. (1978). Part II. Recommended respiratory disease questionnaires for use with adults and children in epidemiologic research: epidemiology standardization project. *Am Rev Respir Dis*, *118*(6, Part II), 7–53.

Fruin, S. A., Winer, A. M. & Rodes, C. E. (2004). Black carbon concentrations in California vehicles and estimation of in-vehicle diesel exhaust particulate mat ter exposures. *Atmos Environ*, *38*(25), 4123–4133.

Gao, P., Jaques, P. A., Hsiao, T., Shepherd, A., Eimer, B. C., Yang, M., Miller, A., Gupta, B. & Shaffer, R. (2011). Evaluation of nano- and submicron particle penetration through ten nonwoven fabrics using a wind-driven approach. *J Occup Environ Hyg*, *8*(1), 13–22.

Gamo, M. (ed) (2011). Risk assessment of manufactured nanomaterials: titanium dioxide (TiO^2). Final report issued on July 22, 2011. NEDO project (P06041) "Research and development of nanoparticle characterisation methods." Japan: New Energy and Industrial Technology Development Organization (NEDO).

Golanski, L., Guiot, A., Rouillon, F., Pocachard, J. & Tardif, F. (2009). Experimental evaluation of personal protection devices against graphite nanoaerosols: fibrous filter media, masks, protective clothing, and gloves. *Hum Exp Toxicol*, *28*(6–7), 353–359.

Golanski, L., Guiot, A. & Tardif, F. (2010). Experimental evaluation of individual protection devices against different types of nanoaerosols: graphite, TiO^2 and Pt. *J Nanopart Res*, *12*(1), 83–89.

Gregoratto, D., Bailey, M. R. & Marsh, J. W. (2010). Modelling particle retention in the alveolarinterstitial region of the human lungs. *J Radiol Prot*, *30*(3), 491–512.

Gregoratto, D., Bailey, M. R. & Marsh, J. W. (2011). Particle clearance in the alveolar-interstitial region of the human lungs: model validation. *Radiat Prot Dosimetry*, *144*(1–4), 353–356.

Grubek-Jaworska, H., Nejman, P., Czumińska, K., Przybyłowski, T., Huczko, A., Lange, H., Bystrzejewski, M., Baranowski, P. & Chazan, R. (2006). Preliminary results on the pathogenic effects of intratracheal exposure to one-dimensional nanocarbons. *Carbon*, *44*(6), 1057–1063.

Han, J. H., Lee, E. J., Lee, J. H., So, K. P., Lee, Y. H., Bae, G. N., Lee, S.-B., Ji, J. H., Cho, M. H. & Yu, I. J. (2008a). Monitoring multiwalled carbon nanotube exposure in carbon nanotube research facility. *Inhal Toxicol*, *20*(8), 741–749.

Han, S. G., Andrews, R., Gairola, G. & Bhalia, D. (2008b). Acute pulmonary effects of combined exposure to carbon nanotubes and ozone in mice. *Inhal Toxicol*, *20*(4), 391–398.

Han, S. G., Andrews, R. & Gairola, C. G. (2010). Acute pulmonary response of mice to multi-wall carbon nanotubes. *Inhal Toxicol*, *22*(4), 340–347.

Harber, P., Conlon, C. & McCunney, R. J. (2003). Occupational medical surveillance. In: McCunney RJ, ed. *A practical approach to occupational and environmental medicine*. Philadelphia, PA: Lippincott, Williams, and Wilkins.

Hinds, W. C. (1999). Respiratory deposition, Chapter 11. In: *Hinds WC, ed. Aerosol Technology: Properties, Behavior, and Measurement of Airborne Particles*, 2nd edition. New York: J. Wiley and Sons.

Hou, P. X., Liu, C. & Cheng, H. M. (2008). Purification of carbon nanotubes. *Carbon*, *15*(46), 2003–3025.

Hubbs, A. F., Mercer, R. R., Coad, J. E., Battelli, L. A., Willard, P., Sriram, K., Wolfarth, M., Castranova, V. & Porter, D. (2009). Persistent pulmonary inflammation, airway mucous metaplasia and migration of multi-walled carbon nanotubes from the lung after subchronic exposure. *The Toxicologist*, *108*, A2193.

Hubbs, A. F., Mercer, R. R., Benkovic, S. A., Harkema, J., Sriram, K., Schwegler-Berry, D., Goravanahally, M. P., Nurkiewicz, T. R., Castranova, V. & Sargent, L. M. (2011). Nanotoxicology—a pathologist's perspective. *Toxicol Pathol*, *39*(2), 301–324.

Huczko, A., Lange, H., Calko, E., Grubek-Jaworska, H. & Droszcz, P. (2001). Physiological testing of carbon nanotubes: are they asbestos-like? *Fullerene Sci Technol*, *9*(2), 251–254.

Huczko, A., Lange, H., Bystrzejewski, M., Baranowski, P., Grubek-Jaworska, H., Nejman, P., Przybyłowski, T., Czumińska, K., Glapiński, J., Walton, D. R. M. & Kroto, H. W. (2005). Pulmonary toxicity of 1-D nanocarbon materials. *Fullerenes, Nanotubes, and Carbon Nanostruct*, *13*(2), 141–145.

IARC. (2010). *IARC monographs on the evaluation of carcinogenic risks to humans: carbon black, titanium dioxide, and talc.*, Vol. *93*. Lyon, France: World Health Organization, International Agency for Research on Cancer. Available at: http://monographs.iarc.fr/ENG/Monographs/vol93/index.php.

ICRP. (1994). Human respiratory tract model for radiological protection. *ICRP publication*, no. *66*. In: Smith H, ed. Annals of the ICRP. Tarrytown, New York: International Commission on Radiological Protection.

Inoue, K., Takano, H., Koike, E., Yanagisawa, R., Sakurai, M., Tasaka, S., Ishizaka, A. & Shimada, A. (2008). Effects of pulmonary exposure to carbon nanotubes on lung and systemic inflammation with coagulatory disturbance induced by lipo- polysaccharide in mice. *Exp Biol Med (Maywood)*, *233*(12), 1583–1590 [Epub ahead of print].

Invernizzi, N. (2011). Nanotechnology between the lab and the shop floor: what are the effects on labor? *J Nanopart Res*, *13*(6), 2249–2268.

IPA. (2013). *Safe handling of polyacrylate dusts*. Prepared by IPA, EDANA, and ASPIA. Contact: CT Helmes, Institute for Polyacrylate Absorbents. Tel 202-296-8120.

ISO. (1995). Air quality—particle size fraction definitions for health-related sampling. Geneva, Switzerland: International Organization for Standardization, ISO Report No. ISO 7708.

ISO/TS. (2008). Nanotechnologies: terminology and definitions for nano-object; nanoparticle, nanofibre and nanoplate. ISO/TS 27687: 2008. Vienna, Austria: International Organization for Standardization.

ISPE. (2007). *Baseline guide*, volume *1*, 2nd edition: Active pharmaceutical ingredients revision to bulk pharmaceutical chemicals [http://www.ispe.org/ baseline-guides/active-pharmaceutical-ingredients].

Johnson, D. R., Methner, M. M., Kennedy, A. J. & Steevens, J. A. (2010). Potential for occupational exposure to engineered carbon-based nanomaterials in environmental laboratory studies. *Environ Health Perspect*, *118*(1), 49–54.

Kagan, V. E., Tyurina, Y. Y., Tyurina, V. A., Konduru, N. V., Potapovich, A. I., Osipov, A. N., Kisin, E. R., Schwegler-Berry, D., Mercer, R., Castranova, V. & Shvedova, A. A. (2006). Direct and indirect effects of single walled carbon nanotubes on RAW 264.7 macrophages: role of iron. *Toxicol Lett*, *165*(1), 88–100.

Kagan, V. E., Konduru, N. V., Feng, W., Allen, B. L., Conroy, J., Volkov, Y., Vlasova, I. I., Belikova, N. A., Yanamala, N., Kapralov, A., Tyurina, Y. Y., Shi, J., Kisin, E. R., Murray, A. R., Franks, J., Stoiz, D., Gou, P., Klein-Seetharaman, J., Fadeel, B., Star, A. & Shvedova, A. A. (2010). Carbon nanotubes degraded by neutrophil myeloperoxidase induce less pulmonary inflammation. *Nat Nanotechnol*, *5*, 354–358.

Kennedy, E. R., Fischbach, T. J., Song, R., Eller, P. M. & Shulman, S. A. (1995). *Guidelines for air sampling and analytical method development and evaluation*. Cincinnati, OH: U.S. Department of Health and Human Services, Centers for Disease Control and Prevention, National Institute for Occupational Safety and Health, DHHS (NIOSH) Publication No. 95–117.

Kim, J. S., Lee, K., Lee, Y. H. , Cho, H. S., Kim, K. H., Choi, K. H., Lee, S. H., Song, K. S., Kang, C. S. & Yu, I. J. (2011). Aspect ratio has no effect on genotoxicity of multi-wall carbon nanotubes. *Arch Toxicol*, *85*(7), 775–786.

Kirchstetter, T. W., Corrigan, C. E. & Novakov, T. (2001). Laboratory and field investigation of the adsorption of gaseous organic compounds onto quartz filters. *Atmos Environ*, *35*(9), 1663–1671.

Kisin, E. R., Murray, A. R., Keane, M. J., Shi, Z. C., Schwegler-Berry, D., Gorelik, O., Arepalli, S., Castranova, V., Wallace, W. E., Kagan, V. E. & Shvedova, A. A. (2007). Single-walled carbon nanotubes: geno- and cytotoxic effects in lung fibroblast V79 cells *J Toxicol Environ Health Part*, *A 70*(24), 2071–2079.

Kisin, E. R., Murray, A. R., Sargent, L., Lowry, D., Chirila, M., Siegrist, K. J., Schwegler-Berry, D., Leonard, S., Castranova, V., Fadeel, B., Kagan, V. E. & Shvedova, A. A.

(2011). Genotoxicity of carbon nanofibers: are they potentially more or less dangerous than carbon nanotubes or asbestos? *Toxicol Appl Pharmacol*, *252*(1), 1–10.

Kobayashi, N., Naya, M., Ema, M., Endoh, S., Maru, J., Mizuno, K. & Nakanishi, J. (2010). Biological response and morphological assessment of individually dispersed multi-wall carbon nanotubes in the lung after intratracheal instillation in rats. *Toxicology*, *276*(3), 143–153.

Kobayashi, N., Naya, M., Mizuno, K., Yamamoto, K., Ema, M. & Nakanishi, J. (2011). Pulmonary and systemic responses of highly pure and well-dispersed single-wall carbon nanotubes after intratracheal instillation in rats. *Inhal Toxicol*, *23*(13), 814–828.

Kromhout, H. (2009). Design of measurement strategies for workplace exposures. *Occup Environ Med*, *59*(5), 349–354.

Ku, B.-K., Emery, M. S., Maynard, A. D., Stolzenburg, M. R. & McMurry, P. H. (2006). In situ structure characterization of airborne carbon nanofibres by a tandem mobility-mass analysis. *Nanotechnology*, *17*, 3613–3621.

Kuempel, E. D. (2000). Comparison of human and rodent lung dosimetry models for particle clearance and retention. *Drug Chem Toxicol*, *23*(1), 203–222.

Kuempel, E. D., O'Flaherty, E. J., Stayner, L. T., Smith, R. J., Green, F. H. Y. & Vallyathan, V. (2001a). A biomathematical model of particle clearance and retention in the lungs of coal miners: Part I. Model development. *Regul Toxicol Pharmacol*, *34*(1), 69–87.

Kuempel, E. D., Tran, C. L., Smith, R. J. & Bailer, A. J. (2001b). A biomathematical model of particle clearance and retention in the lungs of coal miners. Part II. Evaluation of variability and uncertainty. *Reg Toxicol Pharmacol*, *1*(34), 88–101.

Kuempel, E. D., Tran, C. L., Castranova, V. & Bailer, A. J. (2006). Lung dosimetry and risk assessment of nanoparticles: evaluating and extending current models in rats and humans. *Inhal Toxicol*, *18*(10), 717–724.

Kulinowski, K. & Lippy, B. (2011). Training workers on risks of nanotechnology. Washington, DC: National Clearinghouse for Worker Safety and Health Training. National Institute of Environmental Health Sciences (NIEHS) Worker Education and Training Program (WETP) [http://tools.niehs.nih. gov/wetp/index.cfm?id=537].

Kulkarni, P., Sorensen, C. M., Baron, P. A. & Harper, M. (2011). Nonspherical particle measurements: shape factor, fractals, and fibers, In: Kulkarni P, Baron PA, Willeke K, eds. *Aerosol measurement: principles, techniques, and applications*. New York: John Wiley and Sons.

Lam, C. W., James, J. T., McCluskey, R. & Hunter, R. L. (2004). Pulmonary toxicity of single-wall carbon nanotubes in mice 7 and 90 days after intratracheal instillation. *Toxicol Sci*, *77*(1), 126–134.

Lee, J. H., Lee, S. B., Bae, G. N., Jeon, K. S., Yoon, J. U., Ji, J. H., Sung, J. H., Lee, B. G., Lee, J. H., Yang, J. S., Kim, H. Y., Kang, C. S. & Yu, I. J. (2010). Exposure assessment of carbon nanotube manufacturing workplaces. *Inhalation Toxicology*, *22*(5), 369–381.

Legramante, J. M., Valentini, F., Magrini, A., Palleschi, G., Iavicoli, I., Pallante, M., Moscone, D., Galante, A., Bergamashi, E., Bergamashi, A. & Pietroiusti, A. (2009). Cardiac autonomic regulation after lung exposure to carbon nanotubes. *Hum Exp Toxicol*, June (28), 369–375.

Leidel, N. A. & Busch, K. A. (1994). Statistical design and data analysis requirements. Chapter 10. In: Harris RL, Cralley LJ, Cralley LV, eds. *Patty's industrial hygiene and toxicology*, 3rd ed., Vol. *3* Part A. New York: John Wiley and Sons, 453–582.

Li, S., He, P., Dong, J., Guo, Z. & Dai, L. (2005). DNA-directed self-assembling of carbon nanotubes. *J Am Chem Soc*, *127*(1), 14–15.

Li, J. G., Li, W. X., Xu, J. Y., Cai, X. Q., Liu, R. L., Li, Y. J., Zhao, Q. F. & Li, Q. N. (2007). Comparative study of pathological lesions induced by multiwalled carbon nanotubes in lungs of mice by intratracheal instillation and inhalation. *Environ Toxicol*, *22*(4), 415–421.

Li, Z., Hulderman, T., Salmen, R., Chapman, R., Leonard, S. S., Shvedova, A., Luster, M. I. & Simeonova, P. P. (2007). Cardiovascular effects of pulmonary exposure to single-wall carbon nanotubes. *Environ Health Perspect*, *3*(115), 77–82.

Liang, G., Yin, L., Zhang, J., Liu, R., Zhang, T., Ye, B. & Pu, Y. (2010). Effects of subchronic exposure to multi- walled carbon nanotubes in mice. *J Toxicol Environ Health, A 73*(7), 463–470.

Lindberg, H., Falck, G. C., Suhonen, S., Vippola, M., Vanhala, E., Catalan, J., Savolainen, K. & Norppa, H. (2009). Genotoxicity of Nanomaterials: DNA damage and micronuclei induced by carbon nanotubes and graphite nanofibres in human bronchial epithelial cells in vitro. *Toxicol Lett*, *186*(3), 166–173.

Lison, D. & Muller, J. (2008). Lung and systemic responses to carbon nanotubes (CNT) in mice. *Toxicol Sci*, *101*(1), 179–180.

Liu, A., Sun, K., Yang, J. & Zhao, D. (2008). Toxicological effects of multi-wall carbon nanotubes in rats. *J Nanopart Res*, *10*(8), 1303–1307.

Liu, X., Hurt, R. H. & Kane, A. B. (2010). Biodurability of single-walled carbon nanotubes depends on surface functionalization. *Carbon N Y*, *48*(7), 1961–1969.

Lux Research. (2007). The nanotech report. 5th ed. New York: Lux Research.

Lyles, R. H., Kupper, L. L. & Rappaport, S. M. (1997). A lognormal distribution-based exposure assessment method for unbalanced data. *Ann Occup Hyg*, *41*(1), 63–76.

Mader, B. T., Flagan, R. C. & Seinfeld, J. H. (2001). Sampling atmospheric carbonaceous aerosols using a particle trap impactor/denuder sampler. *Environ Sci Technol*, *35*(24), 4857–4867.

Mader, B. T., Schauer, J. J., Seinfeld, J. H., Flagan, R. C., Ye, J. Z., Yang, H., Lim, H. J., Turpin, B. J., Deminter, J. T., Heidemann, G., Bae, M. S., Quinn, P., Bates, T., Eatough, D. J., Huebert, B. J., Bertram, T. & Howell, S. (2003). Sampling methods used for the collection of particle-phase organic and elemental carbon during ACE-Asia. *Atmos Environ*, *37*(11), 1435–1449.

Magrez, A., Kasas, S., Salicio, V., Pasquier, N., Seo, J. W., Celio, M., Catsicas, S., Schwaller, B. & Forró, L. (2006). Cellular toxicity of carbon-based nanomaterials. *Nano Lett*, *6*(6), 1121–1125.

Ma-Hock, L., Treumann, S., Strauss, V., Brill, S., Luizi, F., Mertler, M., Wiench, K., Gamer, A. O., Ravenz- waay, B. & Landsiedel, R. (2009). Inhalation toxicity of multi-wall carbon nanotubes in rats exposed for 3 months. *Toxicol Sci*, *112*(2), 468–481.

Mangum, J. B., Turpin, E. A., Antao-Menezes, A., Cesta, M. F., Bermudez, E. & Bonner, J. C. (2006). Single-walled carbon nanotube (SWCNT)-induced interstitial fibrosis in the lungs of rats is associated with increased levels of PDGF mRNA and the formation of unique intercellular carbon structures that bridge alveolar macrophages in situ. *Part Fibre Toxicol*, *3*(15).

Maynard, A. D. & Kuempel, E. (2005). Airborne nanostructured particles and occupational health. *J Nanopart Res*, 7(6), 587–614.

Maynard, A. D., Baron, P. A., Foley, M., Shvedova, A. A., Kisin, E. R. & Castranova, V. (2004). Exposure to carbon nanotube material: aerosol release during the handling of unrefined single walled carbon nanotube material. *J Toxicol Environ Health, A 67*(1), 87–107.

McDow, S. R. & Huntzicker, J. J. (1990). Vapor adsorption artifact in the sampling of organic aerosol: Face velocity effects. *Atmos Environ, 24A*, 2563–2571.

Mercer, R., Scabilloni, J., Wang, L., Kisin, E., Murray, A. D., Shvedova, A. A. & Castranova, A. V. (2008). Alteration of deposition patterns and pulmonary response as a result of improved dispersion of aspirated single- walled carbon nanotubes in a mouse model. *Am J Physiol Lung Cell Mol Physiol, 294*(1), L87–L97.

Mercer, R. R., Scabilloni, J. F., Wang, L., Battelli, L. A. & Castranova, V. (2009). Use of labeled single walled carbon nanotubes to study translocation from the lungs. *The Toxicologist, 108*, A2192.

Mercer, R. R., Hubbs, A. F., Scabilloni, J. F., Wang, L., Bat-telli, L. A., Schwegler-Berry, D., Castranova, V. & Porter, D. W. (2010). Distribution and persistence of pleural penetrations by multi-walled carbon nanotubes. *Particle Fibre Toxicol, 7*(28), 1–11.

Mercer, R. R., Hubbs, A. F., Scabilloni, J. F., Wang, L., Bat-telli, L. A., Friend, S., Castranova, V. & Porter, D. W. (2011). Pulmonary fibrotic response to aspiration of multi-walled carbon nanotubes. *Part Fibre Toxicol, 8*(1), 21.

Methner, M. M., Birch, M. E., Evans, D. E., Ku, B.-K., Crouch, K. G. & Hoover, M. D. (2007). Mazzuckelli LF, ed. Case study: identification and characterization of potential sources of worker exposure to carbon nanofibers during polymer composite laboratory operations. *J Occup Environ Hyg, 4*(12), D125–D130.

Methner, M., Hodson, L. & Geraci, C. (2010a). Nanoparticle emission assessment technique (NEAT) for the identification and measurement of potential inhalation exposure to engineered nanomaterials—part A. *J Occup Environ Hyg, 7*(3), 127–132 [http://dx.doi. org/10.1080/15459620903476355].

Methner, M., Hodson, L., Dames, A. & Geraci, C. (2010b). Nanoparticle emission assessment technique (NEAT) for the identification and measurement of potential inhalation exposure to engineered nanomaterials— part B: Results from 12 field studies. *J Occup Environ Hyg, 7*(3), 163–176.

Milne, W. I., Mann, M., Dijon, J., Bachmann, P., McLaughlin, J., Robertson, J., Teo, K. B. K., Lewalter, A., de Souza, M., Boggild, P., Briggs, A., Mogensen, K. B., Gabriel, J.-C. P., Roche, S. & Baptist, R. (2008). Carbon nanotubes. *E-Na- no Newsletter, 13*, 5–32 [http://www.phantomsnet. net/files/E_NANO_Newsletter_Issue13.pdf].

Mitchell, L. A., Gao, J., Wal, R. V., Gigliotti, A., Burchiel, S. W. & McDonald, J. D. (2007). Pulmonary and systemic immune response to inhaled multiwalled carbon nanotubes. *Toxicol Sci, 100*(1), 203–214.

Mitchell, L. A., Lauer, F. T., Burchiel, S. W. & McDonald, J. D. (2009). Mechanisms for how inhaled multiwalled carbon nanotubes suppress systemic immune function in mice. *Nat Nanotechnol, 4*(7), 451–456.

Monteiro-Riviere, N. A., Nemanich, R. J., Inman, A. O., Wang, Y. Y. & Riviere, J. E. (2005). Multi-walled carbon nanotubes interaction with human epidermal keratinocytes. *Toxicol Lett, 155*(3), 377–384.

Morimoto, Y., Hirohashi, M., Kobayashi, N., Ogami, A., Horie, M., Oyabu, T., Myojo, T., Hashiba, M., Mizuguchi, Y., Kambara, T., Lee, B. W., Kuroda, E., Shimada, M., Wang,

W. N., Mizuno, K., Yamamoto, K., Fujita, K., Nakanishi, J. & Tanaka, I. (2011). Pulmonary toxicity of well-dispersed single-wall carbon nanotubes after inhalation. *Nanotoxicology*, Sep 26. [Epub ahead of print].

Morimoto, Y., Hirohashi, M., Ogami, A., Oyabu, T., Myojo, T., Todoroki, M., Yamamoto, M., Hashiba, M., Mizuguchi, Y., Lee, B. W., Kuroda, E., Shimada, M., Wang, W. N., Yamamoto, K., Fujita, K., Endoh, S., Uchida, K., Kobayashi, N., Mizuno, K., Inada, M., Tao, H., Nakazato, T., Nakanishi, J. & Tanaka, I. (2012). Pulmonary toxicity of well-dispersed multi-wall carbon nanotubes following inhalation and intratracheal instillation. *Nanotoxicology*, *6*(6), 587–599.

Morrow, P. E. (1988). Possible mechanisms to explain dust overloading of the lungs. *Fundam Appl Toxicol*, *10*(3), 369–384.

Muhle, H., Creutzenberg, O., Bellman, B., Heinrich, U., Ketkar, M. & Mermelstein, R. (1990). Dust overloading of lungs: investigations of various materials, species differences, and irreversibility of effects. *J Aerosol Med*, *3* (Suppl. 3), S111–S128.

Muhle, H., Bellmann, B., Creutzenberg, O., Dasenbrock, C., Ernst, H., Kilpper, R., MacKenzie, J. C., Morrow, P., Mohr, U., Takenaka, S. & Mermelstein, R. (1991). Pulmonary response to toner upon chronic inhalation exposure in rats. *Fund Appl Toxicol*, *17*, 280–299.

Muller, J., Huaux, F., Moreau, N., Misson, P., Heilier, J. F., Delos, M., Arras, M., Fonseca, A., Nagy, J. B. & Lison, D. (2005). Respiratory toxicity of multi-wall carbon nanotubes. *Toxicol Appl Pharmacol*, *207*(3), 221–231.

Muller, J., Huaux, F., Fonseca, A., Nagy, J. B., Moreau, N. & Delos, M. (2008a). Structural defects play a major role in the acute lung toxicity of multiwall carbon nanotubes: toxicological aspects. *Chem Res Toxicol*, *21*(9), 1698–1705.

Muller, J., Decordier, I., Hoet, P. H., Lombaert, N., Thomassen, L., Huaux, F., Lison, D. & Kirsch-Volders, M. (2008b). Clastogenic and aneugenic effects of multi-wall carbon nanotubes in epithelial cells. *Carcinogenesis*, *29*(2), 427–433.

Muller, J., Delos, M., Panin, N., Rabolli, V., Huaux, F. & Lison, D. (2009). Absence of carcinogenic response to multiwall carbon nanotubes in 2-year bioassay in the peritoneal cavity of the rat. *Toxicol Sci*, *110*(2), 442–448.

Murphy, F. A., Poland, C. A., Duffin, R., Al-Jamal, K. T., Ali-Boucetta, H., Nunes, A., Byrne, F., Prina-Mello, A., Volkov, Y., Li, S., Mather, S. J., Bianco, A., Prato, M., Macnee, W., Wallace, W. A., Kostarelos, K. & Donaldson, K. (2011). Length-dependent retention of carbon nanotubes in the pleural space of mice initiates sustained inflammation and progressive fibrosis on the parietal pleura. *Am J Pathol*, *178*(6), 2587–2600.

Murr, L. E., Bang, J. J., Esquivel, E. V., Guerrero, P. A. & Lopez, D. A. (2004a). Carbon nanotubes, nanocrystal forms, and complex nanoparticle aggregates in common fuel-gas combustion sources and the ambient air. *J Nanopart Res*, *6*(2), 241–251.

Murr, L. E., Bang, J. J., Lopez, D. A., Guerrero, P. A. & Esquivel, E. V. (2004b). Carbon nanotubes and nanocrystals in methane combustion and the environmental implications. *J Mater Sci*, *39*(6), 2199–2204.

Murray, A. R., Kisin, E., Leonard, S. S., Young, S. H., Kommineni, C., Kagan, V. E., Castranova, V. & Shvedova, A. A. (2009). Oxidative stress and inflammatory response in dermal toxicity of single-walled carbon nanotubes. *Toxicology*, *257*(3), 161–171.

Murray, A. R., Kisin, E. R., Tkach, A. V., Yanamala, N., Mercer, R., Young, S. H., Fadeel, B., Kagan, V. E. & Shvedova, A. A. (2012). Factoring in agglomeration of carbon nanotubes and nanofibers for better prediction of their toxicity versus asbestos. *Part Fibre Toxicol*, *9*, 10 [http://dx.doi.org/10.1186/1743-8977-9-10].

Nagai, H., Okazaki, Y., Chew, S. H., Misawa, N., Yamashita, Y., Akatsuka, S., et al. (2011). Diameter and rigidity of multiwalled carbon nanotubes are critical factors in mesothelial injury and carcinogenesis. *PNAS Early edition*, 1–9.

Nakanishi, J. (ed) (2011a). Risk assessment of manufactured nanomaterials: "*Approaches"— Overview of approaches and results*. Final report issued on August 17, 2011. NEDO project (P06041) "Research and Development of Nanoparticle Characterization Methods."

Nakanishi, J. (ed) (2011b). Risk assessment of manufactured nanomaterials: carbon nanotubes (CNTs). Final report issued on August 12, 2011. NEDO project (P06041) "Research and development of nanoparticle characterisation methods."

Nanocyl. (2009). *Responsible care and nanomaterials case study Nanocyl*. Presentation at European Responsible Care Conference, Prague. Brussels, Belgium: The European Chemical Industry Council (CEFIC) [http://www.cefic.org/Documents/Responsible Care/04_Nanocyl.pdf].

NanoSafe. (2008). Are conventional protective devices such as fibrous filter media, respirator cartridges, protective clothing and gloves also efficient for nanoaerosols? DR-325/326-200801-1. Brussels, Belgium: European Commission (EC) [http:// www.nanosafe.org/ home/liblocal/docs/Dissemination%20report/DR1_s.pdf].

Nanotech. (2013). Nanotubes hanging in there... *Nanotech Magazine*, Edition 13, 4–5. [www.future- marketsinc.com]

Nikula, K. J., Avila, K. J., Griffith, W. C. & Mauderly, J. L. (1997). Lung tissue responses and sites of particle retention differ between rats and cynomolgus monkeys exposed chronically to diesel exhaust and coal dust. *Fund Appl Toxicol*, *37*, 37–53.

Nikula, K. J., Vallyathan, V., Green, F. H. & Hahn, F. F. (2001). Influence of exposure concentration or dose on the distribution of particulate material in rat and human lungs. *Environ Health Perspect*, *109*(4), 311–318.

NIOSH. (1977). *Occupational exposure sampling strategy manual*. Cincinnati, OH: U.S. Department of Health and Human Services, Centers for Disease Control, National Institute for Occupational Safety and Health, DHHS (NIOSH) Publication No. 77–173.

NIOSH. (1986). *Occupational respiratory diseases*. Merchant JA, ed. Cincinnati, OH: U.S. Department of Health and Human Services, Centers for Disease Control, National Institute for Occupational Safety and Health, DHHS (NIOSH) Publication No. 86–102.

NIOSH. (1994a). Method 7402: asbestos by TEM. In: *NIOSH manual of analytical methods* (supplement issue 2, 8/15/94). 4th ed. Schlecht PC, O'Conner PF, eds. Cincinnati, OH, U.S. Department of Health and Human Services, Centers for Disease Control, National Institute for Occupational Safety and Health, DHHS (NIOSH) Publication No. 2003–154 [www. cdc.gov/niosh/nmam/pdfs/7402.pdf].

NIOSH. (1994b). *NIOSH manual of analytical methods* (NMAM®). Diesel particulate matter (supplement issued 3/15/03). 4th ed. Schlecht PC, O'Conner PF, eds. Cincinnati, OH, U.S. Department of Health and Human Services, Centers for Disease Control, National Institute for Occupational Safety and Health, DHHS (NIOSH) Publiation No. 94–113 [www.cdc.gov/niosh/nmam/].

NIOSH. (1998). Particulates not otherwise regulated, respirable. Method 0600 (supplement issued January 15, 1998). In: *NIOSH manual of analytical methods*. Cincinnati, OH, U.S. Department of Health and Human Services, Public Health Service, Centers for Disease Control and Prevention, National Institute for Occupational Safety and Health, DHHS (NIOSH) Publication No. 94–113 [http://www.cdc.gov/niosh/nmam/pdfs/0600.pdf].

NIOSH. (2002). *Hazard review: health effects of occupational exposure to respirable crystalline silica*. Cincinnati, OH: U.S. Department of Health and Human Services, Centers for Disease Control, National Institute for Occupational Safety and Health, DHHS (NIOSH) Publication No. 2002–129.

NIOSH. (2005). *NIOSH respirator selection logic*. Cincinnati, OH: U.S. Department of Health and Human Services, Centers for Disease Control, National Institute for Occupational Safety and Health, DHHS (NIOSH) Publication No. 2005–100.

NIOSH. (2006). *Criteria for a recommended standard: occupational exposure to refractory ceramic fibers*. Cincinnati, OH: U.S. Department of Health and Human Services, Centers for Disease Control, National Institute for Occupational Safety and Health, DHHS (NIOSH) Publication No. 2006–125.

NIOSH. (2007). *NIOSH pocket guide to chemical hazards and other databases*. Cincinnati, OH: U.S. Department of Health and Human Services, Centers for Disease Control and Prevention, National Institute for Occupational Safety and Health, DHHS (NIOSH) Publication No. 2005–149.

NIOSH. (2009a). *Approaches to safe nanotechnology: managing the health and safety concerns with engineered nanomaterials*. Cincinnati, OH: U.S. Department of Health and Human Services, Centers for Disease Control, National Institute for Occupational Safety and Health, DHHS (NIOSH) Publication No. 2009–125.

NIOSH. (2009b). *NIOSH current intelligence bulletin 60: interim guidance for medical screening and hazard surveillance for workers potentially exposed to engineered nanoparticles*. Cincinnati, OH: U.S. Department of Health and Human Services, Centers for Disease Control and Prevention, National Institute for Occupational Safety and Health, DHHS (NIOSH) Publication No. 2009–116.

NIOSH. (2010a). *NIOSH current intelligence bulletin: occupational exposure to carbon nanotubes and nanofibers (public review draft)*. Cincinnati, OH: U.S. Department of Health and Human Services, Centers for Disease Control, National Institute for Occupational Safety and Health, DHHS (NIOSH) [http://w w w.cdc.gov/nios h/docket/review/docket161A/pdfs/carbonNanotubeCIB_PublicReviewOfDraft.pdf].

NIOSH. (2010b). *A systematic review of the effectiveness of training and education for the protection of workers*. Toronto: Institute for Work & Health. Cincinnati, OH: U.S. Department of Health and Human Services, Centers for Disease Control, National Institute for Occupational Safety and Health, DHHS (NIOSH) Publication No. 2010–127.

NIOSH. (2011a). *Current intelligence bulletin 63: occupational exposure to titanium dioxide*. Cincinnati, OH: U.S. Department of Health and Human Services, Centers for Disease Control, National Institute for Occupational Safety and Health, DHHS (NIOSH) Publication No. 2011–160.

NIOSH. (2011b). *Current intelligence bulletin 64: coal mine dust exposures and associated health outcomes*. A review of information published since 1995. Cincinnati, OH: U.S. Department of Health and Human Services, Centers for Disease Control, National Institute for Occupational Safety and Health, DHHS (NIOSH) Publication No. 2011–172.

NIOSH. (2012). *General safe practices for working with engineered nanomaterials in research laboratories*. Cincinnati, OH: U.S. Department of Health and Human Services, Centers for Disease Control, National Institute for Occupational Safety and Health, DHHS (NIOSH) Publication No. 2012–147.

Noll, J. D. & Birch, M. E. (2008). Effects of sampling artifacts on occupational samples of diesel particulate matter. *Environ Sci Technol*, *42*(14), 5223–5228.

NRC. (2009). Science and decisions: advancing risk assessment. Committee on Improving Risk Analysis Approaches Used by the U.S. EPA, Board on Environmental Studies and Toxicology, Division on Earth and Life Studies, National Research Council of the National Academies. Washington, DC: The National Academies Press.

Nurkiewicz, T. R., Porter, D. W., Barger, M., Millecchia, L., Rao, K. M., Marvar, P. J., Hubbs, A. F. & Castranova, V. (2006). Systemic microvascular dysfunction and inflammation after pulmonary particulate matter exposure. *Environ Health Perspect*, *3*(114), 412–419.

Nurkiewicz, T. R., Porter, D. W., Hubbs, A. F., Stone, S., Chen, B. T., Frazer, D. G., Boegehold, M. A. & Castranova, V. (2009). Pulmonary nanoparticle exposure disrupts systemic microvascular nitric oxide signaling. *Toxicol Sci*, *1*(110), 191–203.

Oberdörster, G., Ferin, J. & Morrow, P. E. (1992). Volumetric loading of alveolar macrophages (AM): A possible basis for diminished AM-mediated particle clearance. *Exp Lung Res*, *18*, 87–104.

Oberdörster, G., Oberdörster, E. & Oberdörster, J. (2005b). Nanotoxicology: an emerging discipline evolving from studies of ultrafine particles. *Environ Health Perspect*, *113*(7), 823–839.

Oyabu, T., Myojo, T., Morimoto, Y., Ogami, A., Hirohashi, M., Yamamoto, M., Todoroki, M., Mizuguchi, Y., Hashiba, M., Lee, B. W., Shimada, M., Wang, W. N., Uchida, K., Endoh, S., Kobayashi, N., Yamamoto, K., Fujita, K., Mizuno, K., Inada, M., Nakazato, T., Nakanishi, J. & Tanaka, I. (2011). Biopersistence of inhaled MWCNT in rat lungs in a 4-week well-characterized exposure. *Inhal Toxicol*, *23*(13), 784–791.

Old, L. & Methner, M. (2008). Effectiveness of local exhaust ventilation (LEV) in controlling engineered nanomaterial emissions during reactor cleanout operations. *J Occup Environ Hyg*, *5*(6), D63–D69.

Olson, D. A. & Norris, G. A. (2005). Sampling artifacts in measurement of elemental organic carbon: Low-volume sampling in indoor and outdoor environments. *Atmos Environ*, *39*(30), 5437–5445.

OSHA, NIOSH. (2011). OSHA NIOSH infosheet: maximize your spirometry screening and surveillance resources. OSHA Publication No. 3415-1-11. Washington, DC: Occupational Safety and Health Administration (OSHA) and Cincinnati, OH: National Institute for Occupational Safety and Health (NIOSH) [http://www.osha.gov/Publications/ osha3415.html].

Osmond-McLeod, M. J., Poland, C. A., Murphy, F., Waddington, L., Morris, H., Hawkins, S. C., Clark, S., Aitken, R., McCall, M. J. & Donaldson, K. (2011). Durability and inflammogenic impact of carbon nanotubes compared with asbestos fibres. *Part Fibre Toxicol*, *8*, 15.

Pacurari, M., Yin, X. J., Zhao, J., Ding, M., Leonard, S. S., Schwegler-Berry, D., Ducatman, B. S., Sbarra, D., Hoover, M. D., Castranova, V. & Vallyathan, V. (2008). Raw Single-wall carbon nanotubes induce oxidative stress and activate MAPKs, AP-1, NF-kappaB, and Akt in normal and malignant human mesothelial cells. *Environ Health Perspect*, *116*(9), 1211–1217.

Pauluhn, J. (2010a). Subchronic 13-week inhalation exposure of rats to multiwalled carbon nanotubes: Toxic effects are determined by density of agglomerate structures, not fibrillar structures. *Toxicol Sci*, *113*(1), 226–242.

Pauluhn, J. (2010b). Multi-walled carbon nanotubes (Baytubes): approach for derivation of occupational exposure limit. Regul Toxicol Pharmacol. [http:// dx.doi.org/10.1016/j.yrtph.2009.12.012].

Piegorsch, W. W. & Bailer, A. F. (2005). Quantitative risk assessment with stimulus-response data. Chapter 4. In: *Analyzing Environmental Data*. Chichester, West Sussex, England: John Wiley and Sons.

Poland, C. A., Duffin, R., Kinloch, I., Maynard, A., Wallace, W. A., Seaton, A., Stone, V., Brown, S., Macnee, W. & Donaldson, K. (2008). Carbon nanotubes introduced into the abdominal cavity of mice show asbestos-like pathogenicity in a pilot study. *Nat Nanotechnol*, *3*(7), 423−428.

Porter, D. W., Sriram, K., Wolfarth, M., Jefferson, A., Schwegler-Berry, D., Andrew, M. & Castranova, V. (2008). A biocompatible medium for nanoparticle dispersion. *Nanotoxicol*, *2*(3), 144–154.

Porter, D. W., Wolfarth, M. G., Chen, B. T., McKinney, W., Hubbs, A. F., Battelli, L., Andrews, M., Frazer, D. & Castranova, V. (2009). Pulmonary toxicity of inhaled multi-walled carbon nanotubes. *The Toxicologist*, *108*, A2193.

Porter, D. W., Hubbs, A. F., Mercer, R. R., Wu, N., Wolfarth, M. G., Sriram, K., Leonard, S. S., Battelli, L., Schwegler- Berry, D., Friend, S., Andrew, M., Chen, B. T., Tsuruoka, S., Endo, M. & Castranova, V. (2010). Mouse pulmonary dose- and time course-responses induced by exposure to multi-walled carbon nanotubes. *Toxicology*, *269*(2–3), 136–147.

Pulskamp, K., Diabaté, S. & Krug, H. F. (2007). Carbon nanotubes show no sign of acute toxicity but induce intracellular reactive oxygen species in dependence on contaminants. *Toxicol Lett*, *168*(1), 58–74.

Raabe, O. G., Al-Bayati, M. A., Teague, S. V. & Rasolt, A. (1988). Regional deposition of inhaled monodisperse coarse and fine aerosol particles in small laboratory animals. *Ann Occup Hyg*, *32*, 53–63.

Ramachandran, G., Ostraat, M., Evans, D. E., Methner, M. M., O'Shaughnessy, P., D'Archy, J., Geraci, C. L., Stevenson, E., Maynard, A. & Rickabaugh, K. (2011). A strategy for assessing workplace exposures to nanomaterials. *JOEH*, *8*(11), 673–685.

Rappaport, S. M., Lyles, R. H. & Kupper, L. L. (1995). An exposure-assessment strategy accounting for within—and between—worker sources of variability. *Ann Occup Hyg*, *39*(4), 469–495.

Rengasamy, S., Verbofsky, R., King, W. P. & Shaffer, R. E. (2007). Nanoparticle penetration through NIOSH- approved N95 filtering-facepiece respirators. *J Int Soc Respir Protect*, *24*, 49–59.

Rengasamy, S., King, W. P., Eimer, B. & Shaffer, R. E. (2008). Filtration performance of NIOSH-approved N95 and P100 filtering-facepiece respirators against 4-30 nanometer size nanoparticles. *J Occup Environ Hyg*, *5*(9), 556–564.

Rengasamy, S., Eimer, B. C. & Shaffer, R. E. (2009). Comparison of nanoparticle filtration performance of NIOSH-approved and CE-marked particulate filtering facepiece respirators. *Ann Occup Hyg*, *53*(2), 117–128.

Rengasamy, S. & Eimer, B. C. (2011). Total inward leakage of nanoparticles through filtering facepiece respirators. *Ann Occup Hyg*, *55*(3), 253–263.

Rom, W. N. & Markowitz, S. (2006). Environmental and Occupational Medicine. 4th ed. Lippincott Williams & Wilkins.

Rouse, J. G., Yang, J., Ryman-Rasmussen, J. P., Barron, A. R. & Monteiro-Riviere, N. A. (2007). Effects of mechanical flexion on the penetration of fullerene amino acid-derivatized peptide nanoparticles through skin. *Nano Lett*, *7*(1), 155–160.

Ryman-Rasmussen, J. P., Riviere, J. E. & Monteiro-Riviere, N. A. (2006). Penetration of intact skin by quantum dots with diverse physicochemical properties. *Toxicol Sci*, *91*(1), 159–165.

Ryman-Rasmussen, J. P., Tewksbury, E.W., Moss, O. R., Cesta, M. E., Wong, B. A. & Bonner, J. C. (2009a). Inhaled multiwalled carbon nanotubes potentiate airway fibrosis in murine allergic asthma, *Am J Respir Cell Mol Biol*, *40*(3), 349–358 (Epub ahead of print).

Ryman-Rasmussen, J. P., Cesta, M. F., Brody, A. R., Shipley-Phillips, J. K., Everitt, J. I., Tewksbury, E. W., Moss, O. R., Wong, B. A., Dodd, D. E., Andersen, M. E. & Bonner, J. C. (2009b). Inhaled carbon nanotubes reach the subpleural tissue in mice. *Nat Nanotechnol*, *4*(11), 747–751. [http://dx.doi.org/10.1038/ NNANO.2009.305].

Sager, T. M., Wolfarth, M., Porter, D., Castranova, V., Wu, N. & Holian, A. (2011). Effect of surface modification on the bioavailability and inflammatory potential of multi-walled carbon nanotubes. *The Toxicologist*, 120, A1178.

Sanchez, V. C., Pietruska, J. R., Miselis, N. R., Hurt, R. H. & Kane, A. B. (2009). Biopersistence and potential adverse health impacts of fibrous nanomaterials: what have we learned from asbestos? *Nanomed Nanobiotechnol*, *5*(1), 511–529.

Sargent, L. M., Shvedova, A. A., Hubbs, A. F., Salisbury, J. L., Benkovic, S. A., Kashon, M. L., Lowry, D. T., Murray, A. R., Kisin, E. R., Friend, S., McKinstry, K. T., Battelli, L. & Reynolds, S. H. (2009). Induction of aneuploidy by single-walled carbon nanotubes. *Environ Mol Mutagen*, *50*(8), 708–717.

Sargent, L. M., Hubbs, A. F., Young, S. H., Kashon, M. L., Dinu, C. Z., Salisbury, J. L., Benkovic, S. A., Lowry, D. T., Murray, A. R., Kisin, E. R., Siegrist, K. J., Battelli, L., Mastovich, J., Sturgeon, J. L., Bunker, K. L., Shvedova, A. A. & Reynolds, S. H. (2011). Single-walled carbon nanotube-induced mitotic disruption. Mutat Res: Genetic Toxicology and Environment, Dec 8. [Epub ahead of print].

Sayes, C. M., Fortner, J., Guo, W., Lyon, D., Boyd, A. M., Ausman, K. D., Tao, Y. J., Sitharaman, B., Wilson, L. J., Hughes, J. B., West, J. L. & Colvin, V. L. (2004). The differential cytotoxicity of water soluble fullerenes. *Nano Lett*, *4*(10), 1881–1887.

Sayes, C. M., Liang, F., Hudson, J. L., Mendez, J., Guo, W., Beach, J. M., Moore, V. C., Doyle, C. D., West, J. L., Billups, W. E., Ausman, K. D. & Colvin, V. L. (2006). Functionalization density dependence of single-walled carbon nanotubes cytotoxicity in vitro. *Toxicol Lett*, *161*(2), 135–142.

Schauer, J. J., Kleeman, M. J., Cass, G. R. & Simoneit, B. R. (1999). Measurement of emissions from air pollution sources. 2. C-1 through C-30 organic compounds from medium duty diesel trucks. *Environ Sci Technol*, *33*(10), 1578–1587.

Schauer, J. J., Mader, B. T., DeMinter, J. T., Heidemann, G., Bae, M. S., Seinfeld, J. H., Flagan, R. C., Cary, R. A., Smith, D., Huebert, B. F., Bertram, T., Howell, S., Kline, J. T., Quinn, P., Bates, T., Turpin, B., Lim, H. J., Yu, J. Z., Yang, H. & Keywood, M. D.

(2003). ACE-Asia intercomparison of a thermal-optical method for the determination of particle-phase organic and elemental carbon. *Environ Sci Technol*, *37*(5), 993–1001.

Schubauer-Berigan, M. K., Dahm, M. M. & Yenchen, M. S. (2011). Engineered carbonaceous nanomaterials manufacturers in the United States: workforce size, characteristics, and feasibility of epidemiologic studies. *J Occup Environ Med*, *53*(Suppl 6), S62–S67.

Schulte, P. A. & Salamanca-Buentello. (2007). Ethical and scientific issues of nanotechnology in the workplace. *Environ Health Perspect*, *115*(1), 5–12.

Schulte, P., Geraci, C., Zumwalde, R., Hoover, M. & Kuempel, E. (2008). Occupational risk management of engineered nanoparticles. *J Occup Environ Hyg*, *5*, 239–249.

Schulte, P. A., Murashov, V., Zumwalde, R., Kuempel, E. D. & Geraci, C. L. (2010). Occupational exposure limits for nanomaterials: state of the art. *J Nanopart Res*, *12*, 1971–1987.

Schulte, P. A., Kuempel, E. D., Zumwalde, R. D., Geraci, C. L., Schubauer-Berigan, M. K., Castranova, V., Hodson, L., Murashov, V., Dahm, M. M. & Ellenbecker, M. (2012). Focused actions to protect carbon nanotube workers. *Am J Ind Med*, *55*(5), 395–411. (Epub 2012 Mar 5).

Shaffer, R. E. & Rengasamy, S. (2009). Respiratory protection against airborne nanoparticles: a review. *J Nanopart Res*, *11*(7), 1661–1672 [http://dx.doi. org/10.1007/s11051-009-9649-3].

Sheesley, R. J., Schauer, J. J., Smith, T. J., Garshick, E., Laden, F., Marr, L. C. & Molina, L. T. (2008). Assessment of diesel particulate matter exposure in the workplace: freight terminals. J Environ Monit [http:// dx.doi.org/10.1039/b715429a].

Shvedova, A. A., Castranova, V., Kisin, E. R., Schwegler-Berry, D., Murray, A. R., Gandelsman, V. Z., Maynard, A. D. & Baron, P. A. (2003). Exposure to carbon nanotube material: assessment of the biological effects of nanotube materials using human keratinocyte cells. *J Toxicol Environ Health A*, *66*(20), 1901–1918.

Shvedova, A. A., Kisin, E. R., Mercer, R., Murray, A. R., Johnson, V. J., Potapovich, A. I., Tyurina, Y. Y., Gorelik, O., Arepalli, S., Schwegler-Berry, D., Hubbs, A. F., Antonini, J., Evans, D. E., Ku, B.-K., Ramsey, D., Maynard, A., Kagan, V. E., Castranova, V. & Baron, P. (2005). Unusual inflammatory and fibrogenic pulmonary responses to single-walled carbon nanotubes in mice. *Am J Physiol Lung Cell Mol Physiol*, *289*(5), L698–L708.

Shvedova, A. A., Sager, T., Murray, A., Kisin, E., Porter, D. W., Leonard, S. S., Schwegler-Berry, D., Robinson, V. & Castranova, V. (2007). Critical issues in the evaluation of possible effects resulting from airborne nanoparticles. In: Monteiro-Riviere N, Tran L, eds. *Nanotechnology: characterization, dosing, and health effects*. Chapter 14. Philadelphia: Informa Healthcare, 221–232.

Shvedova, A. A., Kisin, E., Murray, A. R., Johnson, V. J., Gorelik, O., Arepalli, S., Hubbs, A. F., Mercer, R. R., Keo- having, P., Sussman, N., Jin, J., Stone, S., Chen, B., Deye, G., Maynard, A., Castranova, V., Baron, P. A. & Kagan, V. (2008). Inhalation versus aspiration of single walled carbon nanotubes in C57BL/6 mice: inflammation, fibrosis, oxidative stress and mutagenesis. *Am J Physiol Lung Cell Mol Physiol*, *295*(4), L552−L565.

Snipes, M. B. (1989). Longterm retention and clearance of particles inhaled by mammalian species. *Crit Rev Toxicol*, 20(3), 175–211.

Sriram, K., Porter, D., Tsuruoka, S., Endo, M., Jefferson, A., Wolfarth, W., Rogers, G. M., Castranova, V. & Luster, M. I. (2007). Neuroinflammatory responses following exposure to engineered nanomaterials. *The Toxicologist, 96(*1), A1390.

Sriram, K., Porter, D. W., Jefferson, A. M., Lin, G. X., Wolfarth, M. G., Chen, B. T., McKinney, W., Frazer, D. G. & Castranova, V. (2009). Neuroinflammation and blood-brain barrier changes following exposure to engineered nanomaterials. *The Toxicologist, 108*(1), A2197.

Stapleton, P. A., Minarchick, V., Cumpston, A., McKinney, W., Chen, B. T., Frazer, D., Castranova, V. & Nurkiewicz, T. R. (2011). Time-course of improved coronary arteriolar endothelium-dependent dilation after multi-walled carbon nanotube inhalation. *The Toxicologist*, (*120*), A194.

Stockstill, B. L., Chang, L. Y., Ménache, M. G., Mellick, P. W., Mercer, R. R. & Crapo, J. D. (1995). Bronchiolarized metaplasia and interstitial fibrosis in rat lungs chronically exposed to high ambient levels of ozone. *Toxicol Appl Pharmacol, 134*(2), 251–263.

Stone, K. C., Mercer, R. R., Freeman, B. A., Chang, L. Y. & Crapo, J. D. (1992). Distribution of lung cell numbers and volumes between alveolar and nonalveolar tissue. *Am Rev Respir Dis, 146*(2), 454–456.

Stueckle, T. A., Mirshra, A., Derk, R., Rojanasakul, Y., Castranova, V. & Wang, L. (2011). In vitro assessment of potential tumorigenicity of chronic SWCNT and MWCNT exposure to lung epithelium. *The Toxicologist, 120*, A1182.

Sturm, R. & Hofmann, W. (2009). A theoretical approach to the deposition and clearance of fibers with variable size in the human respiratory tract. *J Hazard Mater, 170*(1), 210–218.

Subramanian, R., Khlystov, A. Y., Cabada, J. C. & Robinson, A. L. (2004). Positive and negative artifacts in particulate organic carbon measurements with denuded and undenuded sampling configurations. *Aerosol Sci Technol, 38*, 27–48.

Tabet, L., Bussy, C., Setyan, A., Simon-Deckers, A., Rossi, M. J., Boczkowski, J. & Lanone, S. (2011). Coating carbon nanotubes with a polystyrene-based polymer protects against pulmonary toxicity. *Part Fibre Toxicol, 8*, 3.

Takagi, A., Hirose, A., Nishimura, T., Fukumori, N., Ogata, A., Ohashi, N., Kitajima, S. & Kanno, J. (2008). Induction of mesothelioma in p53+/− mouse by intraperitoneal application of multi-wall carbon nanotube. *J Toxicol Sci, 33*(1), 105–116.

Thostenson, E. T., Ren, Z. F. & Chou, T. W. (2001). Advance in the science and technology of carbon nanotubes and their composites: a review. *Compos Sci Tech, 61*(13), 1899–1912.

Tran, C. L. & Buchanan, D. (2000). Development of a biomathematical lung model to describe the exposure-dose relationship for inhaled dust among U.K. coal miners. Institute of Occupational Medicine, Edinburgh, U.K. IOM Research Report TM/00/02.

Tran, C. L., Jones, A. D., Cullen, R. T. & Donaldson, K. (1999). Mathematical modeling of the retention and clearance of low-toxicity particles in the lung. *Inhal Toxicol, 11*(12), 1059–1076.

Tran, C. L., Buchanan, D., Cullen, R. T., Searl, A., Jones, A. D. & Donaldson, K. (2000). Inhalation of poorly soluble particles. II. Influence of particle surface area on inflammation and clearance. *Inhal Toxicol, 12*, 1113–1126.

Trout, D. B. & Schulte, P. A. (2009). Medical surveillance, exposure registries, and epidemiologic research for workers exposed to nanomaterials. *Toxicology, 269*(2–3), 128–135.

Tsai, S., Hofmann, M., Hallock, M., Ada, E., Kong, J. & Ellenbecker, M. (2009). Characterization and evaluation of nanoparticle release during the synthesis of single-walled and multiwalled carbon nanotubes by chemical vapor deposition. *Environ Sci Technol*, *43*(15), 6017–6023.

Tsai, S., Huang, R. F. & Ellenbecker, M. J. (2010). Airborne nanoparticle exposures while using constant-flow, constant-velocity, and air-curtain isolated fume hoods. *Ann Occup Hyg*, *54*(1), 78–87.

Turpin, B. J., Huntzicker, J. J. & Hering, S. V. (1994). Investigation of organic aerosol sampling artifacts in the Los Angeles basin. *Atmos Environ*, *28*(19), 3061–3071.

Turpin, B. J., Saxena, P. & Andrews, E. (2000). Measuring and simulation particulate organics in the atmosphere: problems and prospects. *Atmos Environ*, *34*(18), 2983–3013.

US EPA. (1988). Reference physiological parameters in pharmacokinetic modeling. Washington, DC: Office of Health and Environment Assessment, Exposure Assessment Group, U.S. Environmental Protection Agency. EPA report no. EPA/600/6-88/004.

US EPA. (1994). Methods for derivation of inhalation reference concentrations and application of inhalation dosimetry. Research Triangle Park, NC: U.S. Environmental Protection Agency. EPA report no. EPA/600/8-90/066F.

US EPA. (1996). Air quality criteria for ozone and related photochemical oxidants. Washington, DC: Of f ice of Res earch and De velopment, National C enter for Environmental Ass essment, U.S. Environmental Protection Agency. EPA report no. EPA/600/P-93/004aF. Vol. *III*, p. 8–78.

US EPA. (2006). Approaches for the application of physiologically based pharmacokinetic (PBPK) models and supporting data in risk assessment. Washington, DC: National Center for Environmental Assessment, Office of Research and Development, U.S. Environmental Protection Agency. [http://cfpub.epa.gov/ncea/cfm/record-isplay.cfm?deid=157668].

US EPA. (2010). Benchmark dose software, Version 2.1.2. Washington, DC: U.S. Environmental Protection Agency, National Center for Environmental Assessment.

US EPA. (2012). Benchmark dose technical guidance. Washington, DC: U.S. Environmental Protection Agency. EPA/100/R-12/001.

US DOE. (2008). Approach to nanomaterial ES & H. Washington, DC: U.S. Department of Energy, U.S. Department of Energy's Nanoscale Science Research Centers. [http://science.energy.gov/~/media/bes/pdf/doe_nsrc_approach_to_nanomaterial_esh.pdf]

Vankoningsloo, S., Piret, J. P., Saout, C., Noel, F., Mejia, J., Zouboulis, C. C., Delhalle, J., Lucas, S. & Toussaint, O. (2010). Cytotoxicity of multi-walled carbon nanotubes in three skin cellular models: effects of sonication, dispersive agents and corneous layer of reconstructed epidermis. *Nanotoxicol*, *4*(1), 84–97.

Varga, C. & Szendi, K. (2010). Carbon nanotubes induce granulomas but not mesotheliomas. *In Vivo*, *24*(2), 153–156.

Wagner, G. R. & Fine, L. J. (2008). Surveillance and health screening in occupational health. In: Wallace RB, ed. Maxcy-Rosenau-Last Public Health and Preventive Medicine. 15th ed. New York: McGraw-Hill Medical Publishing, 759–793.

Walters, A., Ericson, L. M., Casavant, M. J., Liu, J., Colbert, D. T., Smith, K. A. & Smalley, R. E. (1999). Elastic strain of freely suspended single-walled carbon nanotube ropes. *Appl Phys Lett*, *74*, 3803. [http:// dx.doi.org/10.1063/1.124185].

Wang, L., Castranova, V., Mishra, A., Chen, B., Mercer, R. R., Schwegler-Berry, D. & Rojanasakul, Y. (2010a). Dispersion of single-walled carbon nanotubes by a natural lung surfactant for pulmonary in vitro and in vivo toxicity studies. *Part Fibre Toxicol*, *7*, 31.

Wang, L., Mercer, R. R., Rojanasakul, Y. A., Qiu, A., Lu, Y., Scabilloni, J. F., Wu, N. & Castranova, V. (2010b). Direct fibrogenic effects of dispersed single-walled carbon nanotubes on human lung fibroblasts. *J Toxicol Environ Health Part A*, *73*(5), 410–422.

Wang, L., Luanpitpong, S., Castranova, V., Tse, W., Lu, Y., Pongrakhananon, V. & Rojanasakul, Y. (2011). Carbon nanotubes induce malignant transformation and tumorigenesis of human lung epithelial cells. *Nano Lett*, *11*(7), 2796–2803. [Epub ahead of print] [http://dx.doi. org/10.1021/nl2011214]

Wang, x., Xia, T., Ntim, S. A., Ji, Z., Lim, S., Meng, H., Chung, C. M., George, S., Zhang, H., Wang, M., Li, N., Yang, Y., Castranova, V., Mitra, S., Bonner, J. C. & Nel, A. E. (2012). The dispersal state of multi-walled carbon nanotubes elicits pro-fibrogenic cellular response that correlate with fibrogenesis biomarkers and fibrosis in the murine lung. ACS Nano.

Warheit, D. B., Laurence, B. R., Reed, K. L., Roach, D. H., Reynolds, G. A. M. & Webb, T. R. (2004). Comparative pulmonary toxicity assessment of single-wall carbon nanotubes in rats. *Toxicol Sci*, *77*(1), 117–25.

WHO. (2005). Chemical-specific adjustment factors for interspecies differences and human variability: Guidance document for use of data in dose/concentration-response assessment. Harmonization Project Document No. 2. Geneva: World Health Organization.

Wirnitzer, U., Herbold, B., Voetz, M. & Ragot, J. (2009). Studies on the in vitro genotoxicity of baytubes, agglomerates of engineered multi-walled carbonnanotubes (MWCNT). *Toxicol Lett*, *186*(3), 160–165.

Wolfarth, M. G., Porter, D. W., Hubbs, A. F., Leonard, S., Battelli, L., Andrews, M. & Castranova, V. (2009). Pulmonary toxicity of multi-walled carbon nanotubes. *The Toxicologist*, *108*, A2196

Wolfarth, M. G., Chen, B. T., Castranova, V. & Porter, D. W. (2011). Acute pulmonary responses to MWCNT inhalation. *The Toxicologist*, *120*, A53.

Xu, J., Futakuchi, M., Shimizu, H., Alexander, D. B., Yanagihara, K., Fukamachi, K., Suzui, M., Kanno, J., Hirose, A., Ogata, A., Sakamoto, Y., Nakae, D., Omon, T. & Tsuda, H. (2012). Multi-walled carbon nanotubes translocate into the pleural and induce visceral mesothelial proliferation in rats. *Cancer Science*, *103*(12), 2045–2050.

Yamashita, K., Yoshioka, Y., Higashisaka, K., Morishita, Y., Yoshida, T., Fujimura, M., Kayamuro, H., Nabeshi, H., Yamashita, T., Nagano, K., Abe, Y., Kamada, H., Kawai, Y., Mayumi, T., Yoshikawa, T., Itoh, N., Tsunoda, S. & Tsutsumi, Y. (2010). Carbon nanotubes elicit DNA damage and inflammatory response relative to their size and shape. *Inflammation*, *33*(4), 276–280.

Yan, L., Zhao, F., Li, S., Hu, Z. & Zhao, Y. (2011). Low-toxic and safe nanomaterials by surface-chemical design, carbon nanotubes, fullerenes, metallofullerenes, and graphenes. *Nanoscale*, *3*, 362–382.

Yu, M.-F., Louire, O., Dyer, M. J., Moloni, K., Kelly, T. F. & Ruoff, R. S. (2000). The strength and breaking mechanism of multiwalled carbon nanotubes under tensile load. *Science*, *287*(5453), 637–640.

In: Carbon Nanotubes and Nanofibers
Editor: Miriam A. Summerfield
ISBN: 978-1-63117-629-6

Chapter 2

TRAINING WORKERS ON RISKS OF NANOTECHNOLOGY*

Kristen Kulinowski and Bruce Lippy

INTRODUCTION

Joseph "Chip" Hughes, Director of the Worker Education and Training Program of the National Institute of Environmental Health Sciences

Since 1987, the Worker Education and Training Program has been a major national force for developing effective occupational safety and health training. The program has generated guidance on training hazardous waste workers under OSHA's 1910.120 Hazardous Waste Operations and Emergency Response (HAZWOPER) Standard, on using advanced training technologies to train workers, and on training workers exposed to mold. These guidance documents advocate peer training of workers and good use of adult learning techniques.

In the Fall of 2009, the program held a technical meeting entitled, *Global Safety and Health Issues and their Impact on Worker Training*. A panel that included Drs. Kristen Kulinowski from Rice University, Richard Niemeier from the National Institute for Occupational Safety and Health, and Sam Paik from Lawrence Livermore National Lab addressed nanotechnology and control banding and the implication for worker training. The panel was moderated by Dr. Bruce Lippy.

This paper is a direct result of the conference. Rather than produce our customary conference report, we felt the subject of training workers about the risks and benefits of nanomaterials was so important and unexamined that we should explore it in-depth in a discussion paper that can hopefully lead to a guidance document with broad value to workers and employers. To that end, we welcome your comments. Unfortunately, the history of introducing new materials like asbestos, lead, acrylonitrile and polychlorinated biphenyls has been a legacy of government actions *after* disease has been discovered in a worker

* This is an edited, reformatted and augmented version of a February 2011 report from the National Clearinghouse for Worker Safety and Health Training, with funds provided by the National Institute of Environmental Health Sciences.

population. The U.S. government has learned from history and has been much more proactive in identifying the hazards of nanomaterials. The efforts to collect and disseminate nano-related information through outlets like Rice University's International Council on Nanotechnology, the National Nanotechnology Initiative of the federal government that coordinates the activities of 25 Federal agencies, and the Woodrow Wilson International Center for Scholars, has been unprecedented.

The growth of nanotechnologies is so explosive, however, that regulatory agencies are having difficulties keeping up. As the Government Accountability Office (GAO) starkly noted in its May 2010 report on nanotechnology, "Although the body of research related to nanomaterials is growing, the current understanding of the risks posed by these materials is limited." The area that has received the least amount of attention is the training of the nanotechnology workforce. Hopefully, this document will begin to redress that deficit.

1. Purpose and Overview

1.1. Purpose of This Document

Among the thousands of papers published on nanotechnology, this is one of the first to address the critical issue of how workers who are creating and handling nanomaterials should be trained about the hazards they face – in laboratories, manufacturing facilities, at hazardous waste cleanup sites and during emergency responses. Given the limits in the current understanding of nanotoxicology, workplace exposures and effectiveness of control strategies, defining effective training is particularly problematic. But workers clearly have the potential to be exposed and are guaranteed the right to know about the risks they face, under international hazard communication standards that are being harmonized across the globe.

The purposes of this document are to:

1. Provide a broad overview of the key issues that workers and instructors should understand about nanotechnology, particularly the occupational health and safety issues;
2. Define current knowledge on worker protection through work practices, engineering controls and personal protective equipment;
3. Review the applicability of current U.S. regulations to nanomaterials with a focus on the OSHA Hazardous Waste Operations and Emergency Response standard (HAZWOPER);
4. Provide a suggested outline for an 8-hour awareness course that would prove beneficial to most workers handling nanomaterials and may satisfy OSHA's requirement for HAZWOPER refresher training; and
5. Provide a framework that will eventually enable the hundreds of instructors funded by the National Institute of Environmental Health Sciences' Worker Education and Training Program (WETP) to train their constituents about the hazards of and controls for specific nanomaterials in their workplaces. Nearly two million workers have received NIEHS WETP-supported safety and health training since the inception of the program in 1987. The primary objective of the program is to fund non-profit

organizations to provide high quality occupational safety and health training to workers who are involved in handling hazardous materials or in responding to emergency releases of hazardous materials. These are covered by OSHA's Hazardous Waste Operations and Emergency Response (HAZWOPER) standard, 29 CFR 1910.120, which arguably covers nanoparticles.

The term nanotechnology describes a wide range of technologies, materials and applications that are affecting or will affect every sector of commerce including medicine, energy, construction, environmental remediation, automotive and aerospace. Examples that exist today include photocatalytic particles that break down organic pollutants in contaminated groundwater, novel medical devices that demonstrate greater specificity for cancer cells, and fibers that improve mechanical strength while reducing mass in automobile parts. The aspects of nanotechnology that merit the attention of the worker training community are its impact across sectors, its novelty and its potential for growth.

Not a single technology itself, nanotechnology has enjoyed widespread support from the federal government and large investments by industry because it offers a platform for improving existing materials, devices and drugs by exploiting the novel properties that emerge when matter is taken down (or up) to the nanoscale. These new modes of action can significantly enhance the properties of the products in which they are used, leading to materials that are stronger, multifunctional or more energy efficient. But novelty is a double-edged sword; i.e., the same unique properties that benefit a particular application could result in new risks to people or the environment. Concerns about health and safety are being addressed in research laboratories around the world but the application of this risk-relevant research to worker safety is still in the early stages. In science fiction, depictions of nanotechnology abound, making it seem like something from the future; however, there are over a thousand products already on the market today and many more in the pipeline. If the scale of the potentially affected workforce tracks the rapid growth of nanotechnology research and product development, the implications for worker training should be clear: many workers should already be trained in safe handling of nanomaterials and many more will need training in the near future, particularly the secondary and tertiary users.

1.2. Outline

This paper will provide a basic understanding of nanotechnology and its implications for worker training. Section 1 introduces essential concepts of nanotechnology, presents the major areas of its application and identifies unresolved issues around workforce identification. Section 2 reviews the relevant environmental, health and safety literature and what this means for worker training. Section 3 assesses the strengths and weaknesses of applying traditional risk management approaches to the nanotechnology workplace. Regulatory and voluntary approaches developed specifically for use with nanoparticles are described in Section 4. Freely available resources that will help trainers stay up to date on new developments in nanotechnology and worker training are catalogued in Section 5. The paper concludes in Section 6 with concrete suggestions for implementing a nanotechnology-specific training program.

2. INTRODUCTION TO NANOTECHNOLOGY AND NANOPARTICLES

2.1. Definitions

Nanotechnology is an emerging area of technology development involving structures that measure between 1–100 nanometers (nm) in one or more dimensions. While precise definitions are still somewhat variable, most standard definitions recognize that nanotechnology involves science and engineering of matter at the nanoscale where properties may change with size or new properties may emerge. Nanotechnology can be further subdivided into nanomaterials and nanostructured materials. These terms are often used interchangeably but have subtly different meanings depending on whether it is an external dimension or merely a component, internal feature that falls within the nanoscale size range of about 1-100 nm. According to some definitions, **nanomaterials** are small pieces of matter with one or more **external** dimensions on the nanoscale. If the object has two or three external dimensions on the nanoscale it may be referred to as a **nanoparticle**. Quantum dots that measure 4 nm in diameter are nanoparticles. If it has two or three external dimensions on the nanoscale AND has a length-to-width aspect ratio of 3:1 or greater it may be referred to as a **nanofiber**. Nanofibers include carbon nanotubes and nanorods. **Nanostructured materials** have internal features that fall within the nanoscale but may be larger than 100 nm as a whole. Examples of nanostructured materials that are larger than 100 nm include a microscale particle that has nanoscopic internal pores or a 300-nm aggregate of 20-nm primary particles. Nanoparticles are a subset of nanomaterials. For the most part, this paper will focus on nanomaterials but not nanostructured materials.

Smaller than microscale particles, yet larger than atoms and all but the largest molecules, nanoparticles occupy a transitional regime between classical and quantum physics where physical and chemical properties may depend on the nanoparticle's size, structure, composition, surface structure or surface composition. Classical physics governs the behavior of objects in our everyday experience and thus is more intuitively familiar to us. We know, for example, to duck (or put our glove up) when we see a baseball flying toward our head. The ball's trajectory, which is governed by the forces of gravity and friction, is easily predicted and can even be precisely calculated using the principles of classical physics. If that baseball were an object about 10 trillion times smaller, its behavior would be governed by quantum physics and we would lose our ability to predict its location and path. We wouldn't know when to duck. Nanoparticles behave more like quantum objects than like baseballs, especially at the lower end of their size scale. This means that our prior knowledge and experience with a substance may not always predict how that substance will behave when it is made at the nanoscale. As the nanoparticle's size increases, it acts more and more like a baseball, which is why there is an upper boundary on most definitions of the size scale.

Gold affords a great example of the striking differences of materials at the nanoscale. Everyone is aware of the shiny orange-yellow color of gold in coins and jewelry. That is a feature of the macroscale world. Between 100 and about 30 nanometers, gold is purple and at 30 nm in size, a gold particle is bright red. Smaller particles become brownish in color. [1] Macroscale gold is prized for its chemical inertness as well as its luster. But nanoscale gold can be highly reactive, even being used as a catalyst in some chemical reactions. Color, magnetism, electrical conductivity and chemical reactivity are just some of the properties that can change at and throughout the nanoscale.

Nanoparticles can be intentionally designed for a purpose (engineered or manufactured), unintentionally produced as part of another industrial or anthropogenic process (incidental) or produced naturally. Incidental nanoparticles may also be referred to as ultrafine particles. (See Table 1.) To meet the definition of "nanotechnology" the particles must have been engineered; e.g., a microscale titanium dioxide particle may be reduced to the nanoscale to prevent it from appearing white in a sunscreen formulation. In contrast, the nanoparticles that may compose a fraction of diesel exhaust are not intentionally designed to exploit special properties that occur at the nanoscale. Rather they are an accidental by-product of incomplete combustion and therefore are not considered to meet the definition of nanotechnology. Practically speaking the distinction in the mode of production between incidental and intentional nanoparticles may have little bearing on the worker if an exposure leads to unwanted outcomes. However, for the sake of clarity this paper will assume the nanoparticles are engineered.

Engineered nanoparticles can be made from many different chemical substances. The types of nanoparticles in use today can be broadly classified into five categories: metals, metal oxides (ceramics), carbon-based, semiconducting (quantum dots) and organics. (See Table 2). Within these broad classifications, there may be several subcategories, each of which has its own set of properties. A nanoscale metal may be prized for its unique optical, electrical or catalytic properties or, in the case of silver, its antimicrobial activity. Metal oxides may have interesting magnetic, mechanical or catalytic behavior.

Dust plume from the Icelandic volcano, April 17, 2010.

Table 1. Distinguishing nanoparticle types by their mode of production

Nanoparticle Type	Examples
Naturally occurring	Volcanic ash, sea spray, combustion by-product
Incidental (Ultrafine)	Fresh welding fume, freshly generated diesel exhaust
Engineered (Manufactured)	Nanotube, nanoscale titanium dioxide

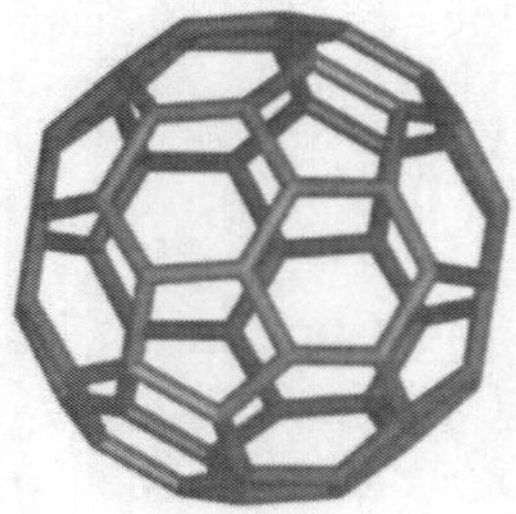

Courtesy Rice University.

Buckminsterfullerene (Buckyball) composed of 60 carbon atoms.

Table 2. Broad categories of nanoparticles

Broad category	Examples
Metals	Silver, Gold, Copper
Metal oxides (ceramics)	Titanium dioxide, Zinc oxide, Cerium oxide
Carbon-based	Fullerenes, Nanotubes
Semiconducting (quantum dots)	Cadmium selenide, Cadmium sulfide, Zinc sulfide
Organic	Polymer beads, Dendrimers

Courtesy Wikimedia Commons.
Computer simulation of a nanotube.

Courtesy Shaddack, Wikimedia Commons.
Multiwalled carbon nanotubes as a black clumpy powder. 10 gram container, scale in centimeters.

Carbon-based nanoparticles may impart mechanical strength and can be made to conduct electricity. Quantum dots have useful optical properties. Organic nanoparticles are especially useful in medical applications.

NANOTECH FOR ENVIRONMENTAL REMEDIATION

As of February 2009, the EPA's National Priorities List contained 1,255 Superfund hazardous waste sites that are estimated to take up to 35 years and cost up to $250 billion to remediate. Federal agencies looking for ways to do cleanup "quicker, cheaper and better" are investigating catalytic nanoparticles as an attractive alternate to expensive pump-and- treat technologies. Magnetic iron oxide nanoparticles have been shown to bind arsenic irreversibly up to 10 times more effectively than micrometer-sized particles. These particles can be separated from water with magnetic fields, yielding a 99 percent cleanup in laboratory and field tests.

A case study at a manufacturing site in New Jersey where the primary contaminants of concern in ground water were trichloroethylene (TCE) and perchloroethylene (PCE) found that using iron oxide nanoparticles would reduce costs 80 to 90 percent and cut the required time even more dramatically, from an average of 18 years for pump-and-treat on EPA sites to – in one study – a 99% reduction in TCE levels within days of injection.

2.2. Quantifying the Size of the Industry and Affected Workforce

Because nanotechnology is a set of technology platforms that are applied across multiple sectors, it is difficult to quantify the nanotechnology "industry." No North American Industry Classification System (NAICS) code for nanotechnology exists and a company may choose whether or not to disclose that it is engaged in commercial nanotechnology research, manufacture or use at its discretion. Nonetheless there have been some attempts to estimate nanotechnology's impact on the marketplace today, as well as to project how much it is expected to grow. A 2008 report by Lux Research estimated that nanotechnology was used in $147 billion worth of products in 2007 and will impact $2.5 trillion of products by 2015. [The Recession's Ripple Effect on Nanotech] Another analysis puts these same figures at $2.3 billion in 2007 and only $81 billion in 2015. [Nanoscale Materials and Markets 2008-2015] Since there are no standard industry codes or even well established standard definitions for nanotechnology, these analyses may better serve as measures of the potential future growth of nanotechnology than as actual market values.

If the size and scope of the industries involved in nanoparticle production or use are difficult to quantify, it is even harder to collect demographic information on the worker population. Such information would be invaluable in the development of a nanotechnology worker registry that would enable medical surveillance to identify potential problems early. The prospect of identifying the nanotech workforce is less daunting when considering a small-to-medium nanotechnology enterprise, as its core business most likely involves the manufacture or use of engineered nanoparticles. But for many large companies, nanoparticle production or use may constitute a small fraction of the total business and the employees who handle nanoparticles may not be distinguished from workers in other parts of the company. For this class of companies the nanotechnology workforce may not differ substantially in key demographic categories from workers in the chemicals industry. NIOSH, in particular, is interested in gaining a greater understanding of the potentially affected workforce though it asserts, "there is insufficient scientific and medical evidence to recommend the specific medical screening of workers potentially exposed to engineered nanoparticles." [2] NIOSH, however, also recommends the continued use of established medical surveillance approaches and the conduct of hazard surveillance. The European Agency for Safety and Health at Work unambiguously noted that, "At present, there is insufficient information on the number of workers exposed to nanomaterials in the work place or the effects on human health of such exposure." [3]

Recent studies indicated that the current number of workers who are directly involved with nanoparticles is probably relatively low. Although there is limited information about the size and scope of the U.S. worker population, studies done on other industrialized nations with strong nanotechnology research and development programs reveal small percentages of workers engaged in handling nanoparticles. Research published in 2010 indicated that only 0.6% of manufacturing companies in Switzerland used nanomaterials, which led to an estimation of only 1,309 workers across the entire country, an average of 2.5 workers per company handling nanomaterials. An earlier study in England estimated 2,000 workers handling nano-materials in that country, although the researcher's definition of a "nano" company was quite proscribed. [4] One market analysis estimates that the global workforce in nanotechnology industries will reach 2 million by the year 2018. [5] A better accounting of the potentially exposed workforce is needed to guide risk management and worker training programs.

As manufacture and use constitute only part of the lifecycle of an engineered nanoparticle, chemical industry workers are only a part of the potentially affected workforce. In addition to production line workers, maintenance and housekeeping staff must be considered. Moreover, smaller pockets of workers will also be found throughout the scores of institutes and universities performing research on novel nanoparticles. These workers may not account for a significant population in any given workplace, but in the aggregate may account for a substantial number of people.

Except at the point of original manufacture, most workers will not encounter nanoparticles in their native form. Instead the worker may encounter a nanoparticle that has been physically or chemically modified for use in an end-product. As an example, carbon nanofibers may be manufactured by one company, sold in their native form to another company, incorporated by that second company into a car bumper, sold in the car bumpers to an auto manufacturer which then sells the car to the final consumer. To measure the full impact of engineered nanoparticles, the entire lifecycle must be assessed, from production through formulation and use to disposal or recycling. The maintenance and custodial workers at these locations may be exposed during their work on labs and systems. Health and safety of workers handling nano-enabled products or materials at their end-of-life must also be considered. The majority of nanomaterials encountered will be various products that are improved or "nano-enabled" by having engineered nanomaterials in them. In some cases, the nanoparticle will persist largely unchanged, whereas in others it will change drastically or even dissolve altogether. Once in the natural environment a nanoparticle may be transported beyond the point of release through waters or soils in unexpected ways, be broken down by microbes or be chemically transformed by oxygen and sunlight. At this point, few complete assessments have been attempted and there is not enough information to predict how a given nanoparticle in a specific application will behave throughout its lifecycle. This results in significant knowledge gaps that can impact the training of workers who handle nano-enabled products in the waste and recycling industries.

It is clear, however, that nanoparticles will figure much more prominently in the cleanup of hazardous waste sites into the future. The NIEHS Worker Education and Training Program was created under the Superfund Amendments and Reauthorization Act of 1986 to train workers to safely remediate the hundreds of thousands of hazardous waste sites in this country. The work continues. The EPA has been allocated nearly a billion dollars for remediation projects under the American Recovery and Reinvestment Act of 2009. [6] Greater than 80 percent of NPL sites have contaminated groundwater. Nanoparticles, like nano zero- valent iron (nZVI), have been shown to be particularly effective in groundwater remediation, which has a direct impact on greater than half of the U.S. population relying on groundwater for drinking. Polluted groundwater cleanup, even though the trend is moving from pump-and-treat to *in situ* methods, has proven to be "protracted, costly, and sometimes infeasible." [7] Nanomaterials like zeolites, metal oxides, carbon nanotubes, bimetallic nanoparticles (BNPs) and titanium dioxide have been successfully applied in remediation, but the use of nZVI dominates and is "increasingly rapidly." [8] See figure 1.

Based on findings of 80 to 90 percent cost reductions using nZVI rather than pump-and-treat for removing chemicals from groundwater noted previously, the EPA projects that using nanoremediation could potentially save this nation $87 to $98 billion on remediating hazardous waste sites over the next 30 years. [9] Consequently, the exposure of hazardous waste workers to nanoparticles will increase. There is, unfortunately, much greater agreement on the effectiveness of treatment than the risk posed by release of particles in the environment. As the Royal Commission on Environmental Pollution noted in 2008:

> "While there have been no significant events that would lead us to suppose that the contemporary introduction of novel materials is a source of environmental hazard, we are acutely aware of past instances where new chemicals and products, originally thought to be entirely benign, turned out to have very high environmental and public health costs." [10]

The Royal Commission primarily addressed acute exposures and release events, which doesn't consider the potential for chronic effects.

2.3. Nanoparticles' Environmental, Health and Safety Impacts

Nanotechnology environmental, health and safety (nano-EHS) research is still in an early phase with published findings scattered across dozens of different journals. Publications of relevance to this topic are collected, catalogued and indexed within the ICON Virtual Journal of Nano-EHS, an open web-based resource that contains citations to more than 4,600 papers. [http://iconThe Nano-EHS Database Analysis Tool permits users to sort this comprehensive resource by particle type, exposure target population, exposure pathway and other criteria. A search on all papers reveals a rapid acceleration in the pace of knowledge generation in the fields of toxicology and environmental impact within the last decade. (See Figure 2.) Reports of direct relevance to worker health, such as research findings on the ability of respirators or gloves to filter out nanoparticles, the ability of local exhaust ventilation to control the flow of nanoparticles or the proper methods of exposure monitoring in the workplace, constitute a very small fraction of this literature.

Despite the large number of papers on the general subject of nano-EHS, it is difficult to draw robust conclusions about the risks engineered nanoparticles might pose to workers or the environment. The reasons for this are myriad, and include a lack of validated protocols for performing toxicology tests on nanoparticles, questions about the appropriate metric for measuring dose, lack of models for how nanoparticles are transformed in the body or the natural environment and the role of surface area and surface chemistry in controlling bio-interactions. [11] Despite these challenges, the large body of work does permit some general conclusions of relevance to worker safety to be drawn.

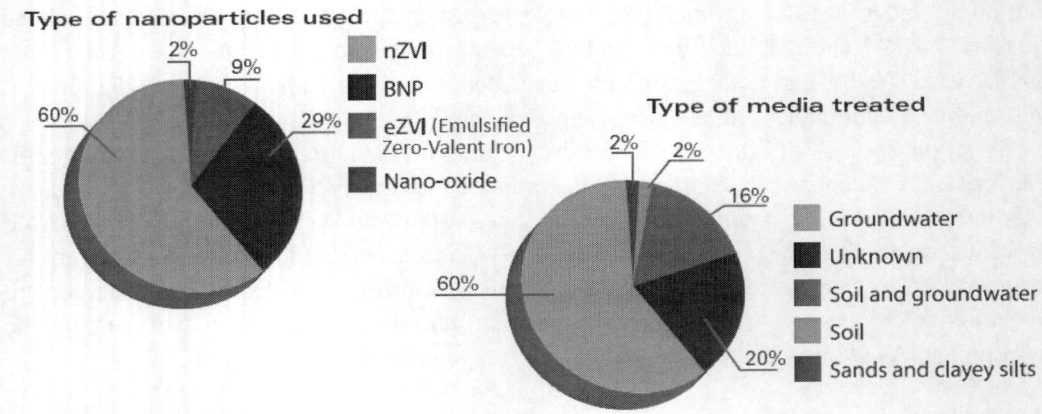

From Karn et al. 2009.

Figure 1. Type of nanoparticles used (A) and type of media treated (B) at 44 international sites.

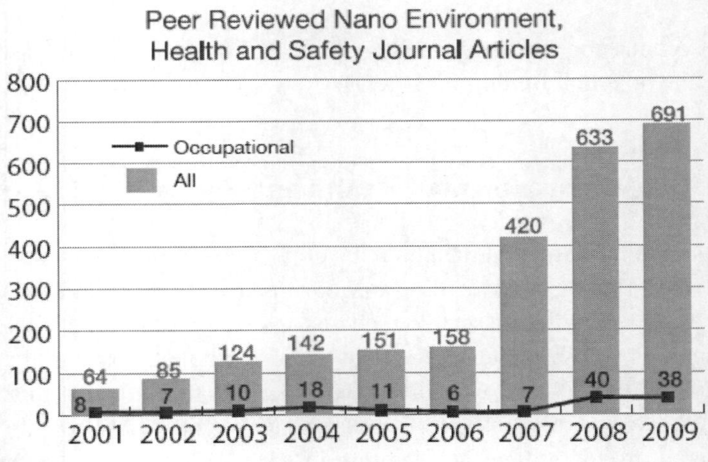

Source: http://icon.

Figure 2. Occupationally Relevant Research. Results from a search of the ICON Virtual Journal on Nanotechnology Environment, Health and Safety for all peer-reviewed nano-EHS papers published between 2000-2009 (green) and those papers that address some aspect of worker safety (purple).

Because of their small size and active surface chemistry, nanoparticles may behave in different ways in the body than their non-nanoscale analogs. The majority of nano-EHS papers address some aspect of hazard, mostly acute toxicity tests done in cell culture. [12] Certain nanoparticles have been shown in animal studies to translocate along the olfactory nerve into the brain, cross the placenta and penetrate damaged or diseased skin. Once inside the body, certain nanoparticles have induced inflammatory responses, cardiovascular effects, pulmonary fibrosis and genotoxicity. Certain carbon nanotubes, one of the most widely researched class of nanoparticles from both a technological and toxicological perspective, have even been shown to induce asbestos-like effects in rodents which raises concerns among occupational safety professionals. [13] It must be emphasized, however, that effects demonstrated by one type of nanoparticle in one laboratory study cannot be generalized to other nanoparticles.

In one highly publicized carbon nanotube study, for example, only the long, straight, multiwalled forms acted in a manner similar to asbestos fibers when injected into the rodents' bodies. Other nanotube forms that were shorter and more flexible did not induce the harmful response, nor have other types of non-carbon-based nanotubes been shown to mimic toxic asbestos fibers. [14] The most recent draft guidance from NIOSH, their Current Intelligence Bulletin, "Occupational Exposure to Carbon Nanotubes and Nanofibers," indicates that mesothelioma has been produced in a strain of mice with multiwalled nanotubes. [15] Examples such as these serve to illustrate not that "nanoparticles are toxic" but that hazard is related to specific material properties such as composition, form, dimension and specific use scenarios.

The research community is working to develop a better understanding of how nanoparticles' physical and chemical characteristics can be correlated to their biological interactions. Without predictive models for linking measurable properties such as size, shape and surface area to biological interactions such as the production of reactive oxygen species, protein misfolding, cell death (apoptosis) and mutagenicity, the number of individual nanoparticle variants that would have to be tested is practically infinite. Experts recently concluded that predictive models are an important long-term goal requiring ten years of work or more. [16]

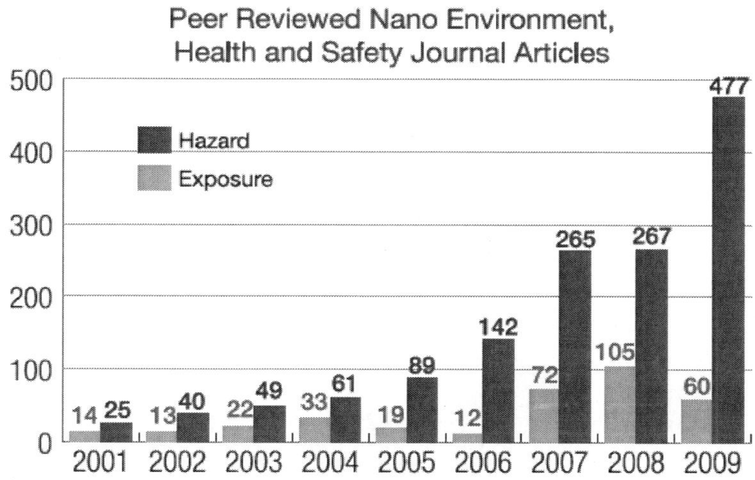

Source: http://icon.rice.edu/report.cfm.

Figure 3. Hazard vs. Exposure. Results from a search of the ICON Virtual Journal on Nanotechnology Environment, Health and Safety for all peer-reviewed nano-EHS papers published between 2000-2009 that address nanomaterial hazards and those papers that address nanomaterial exposure.

The relevance of the nano-EHS hazard studies to real-life working conditions is unclear. Risk assessment relies upon an understanding of both hazard and exposure; when one is absent, the assessment is incomplete. Exposure studies that measure the likelihood that one would come in contact with a toxicant in a high enough dose to cause the unwanted outcome lag far behind the hazard studies as demonstrated in Figure 3. In the absence of nano-specific exposure studies, the likelihood of exposure may be able to be estimated using surrogates, predictors, or historical experience.

3. Application of Traditional Risk Management Approaches to Protect Workers Handling Nanoparticles

Nanoparticles may be new but hazardous substances are not. We need not assume a risk management program has to be constructed from scratch. What is important, however, is to carefully examine our assumptions and validate our existing tools when applying them to situations where nanoparticles are present. Many of the existing frameworks for dealing with toxic substances may be able to be applied to nanoparticles with little revision. Others may need to be scrapped altogether. As always, the first steps are to understand the specific tasks that could bring a worker into contact with a nanoparticle and then identify how best to control that exposure. What follows here is a review of the applicability of existing tools for measuring and controlling exposure to nanoparticles.

3.1. Most Likely Exposures among NIEHS Representative Groups

NIOSH, working with the private sector, has attempted to identify the tasks that are most likely to generate worker exposures to nanoparticles. The following have been specifically noted:

- Generating nanoparticles in the gas phase in non-enclosed facilities;
- Handling nanostructured powders;
- Working with nanoparticles in liquid media without adequate protection, particularly gloves;
- Working with nanoparticles in liquid during pouring or mixing operations or where a high degree of agitation is involved;
- Conducting maintenance on equipment and processes used to produce or fabricate nanoparticles; and
- Cleaning up spills or waste materials. [17]

As noted earlier, the NIEHS Worker Education and Training Program has been funding training for high-risk populations of hazardous waste workers and emergency responders since 1987 under the Superfund Amendment and Reauthorization Act. The organizations that provide this training serve distinct populations that may benefit from training in the identification and control of nanomaterials hazards. Table 3 is the result of an informal survey of principal investigators from the NIEHS-funded programs.

Table 3. Potential exposures among NIEHS-supported worker populations

NIEHS WETP awardee	Worker population	Types of nanomaterials	Tasks
(CPWR) The Center for Construction Research and Training	Workers involved in construction, demolition and remediation	nano-scale zero- valent iron; photocatalytic concrete (which uses nano-scale TiO_2 as an additive); nano-enabled construction materials	Using nano-scale iron to treat soil contaminated with chlorinated solvents; demolishing structures containing photocatalytic concrete
International Association of Fire Fighters	Firefighters	Wide range of nanomaterials	Responding to fires, explosions or leaks at firms producing carbon nanotubes
Teamsters	Truck drivers	Solutions of nanoparticle such as multiwalled carbon nanotubes and packaged dry nanoparticles	Driving trucks carrying packaged nanoparticles that may be involved in a spill on the highway
International Union of Chemical Workers	Chemical plant industrial workers	Wide range of nanoparticles	Producing batches of nanoparticles in chemical plants
International Union of Operating Engineers	Stationary engineers	Nanosilver biocides	Adding biocides to cooling tower water and HVAC drip pans
Laborers International Union of North America	Cleanup workers	Nanoparticles in hazardous waste	Performing cleanup of waste sites
Midwest Consortium	Industrial emergency responders	Wide range used in production of personal care products, paints and new ones daily	Spill cleanup and response
Midwest Consortium	Municipal first- on-scene	Wide range of materials that may be spilled during transport	Spill cleanup and response
Midwest Consortium	Treatment, Storage and Disposal Facilities (TSDF)	Wide range	Transfer, repackaging and disposal
Service Employees International Union	Nurses and physicians; facility maintenance workers, janitorial services	Nanoparticles used in medicines; nano-enabled cleaners and disinfectants	Preparing intravenous liquids
Dillard University	Cleanup workers	Nanomaterials in hazardous waste	Performing cleanup of waste sites
United Steelworkers	DOE cleanup site workers	Nanoremediation agents like Nano iron oxide	Performing in situ injection of agents
United Auto Workers		Nanomaterials in autoparts	Assembling cars

3.2. Assessing Exposures

3.2.1. Difficulty with the Standard IH Paradigm

The standard model of industrial hygiene that has underpinned the profession as well as the OSHA regulatory approach since 1970 has been the measurement of exposures in the personal breathing zone of workers for comparison against established occupational exposure limits. These limits were established based on animal tests and, too often, upon human epidemiology; but given the decades of use of standard industrial solvents like toluene, the accumulated data were often impressive and persuasive for setting a limit that would arguably allow a working career of exposure without permanent, deleterious health effects. These numbers were generally set to be an 8-hour time-weighted average. For dust exposures, the unit of measurement has historically been milligrams (or micrograms) per cubic meter of air.

Given the extraordinarily small size and weight of nanoparticles, however, these mass-based measurements are of limited value. The majority of animal toxicology studies have used mass as the exposure dose for evaluating adverse health outcomes. Until improved sampling and analytical methods can be developed, and until data become available to determine if alternative exposure metric to mass may be more biologically relevant, mass-based measurements may have to be applied along with other exposure metrics for evaluating worker exposures. One critical feature of nanoparticles is the tremendous surface area that is created for the same amount of mass. A thought experiment can illustrate how significant this can be for determining the nanomaterial's behavior. Consider that we have a large block of pure, solid gold that measures 1 meter on each side. This makes a surface area of 6 m_2. Now imagine that we cut that block of gold into pieces that each measure ¼ m on a side; the surface area is now 24 m^2. Taking this thought experiment to the extreme, if we divide the same mass of gold into particles measuring 1 nm on each side, the surface area exceeds 6 billion m^2. That is more than enough to cover the surface of the entire state of Delaware.

This is also of concern for health effects: the same reactive surfaces that are prized in creating unique properties in new products appear to be implicated in much of the unwanted health effects. As the British Standards Institute correctly noted, "Altered chemical and/or physical properties might be expected to be accompanied by altered biological properties, some of which could imply increased toxicity." [18]

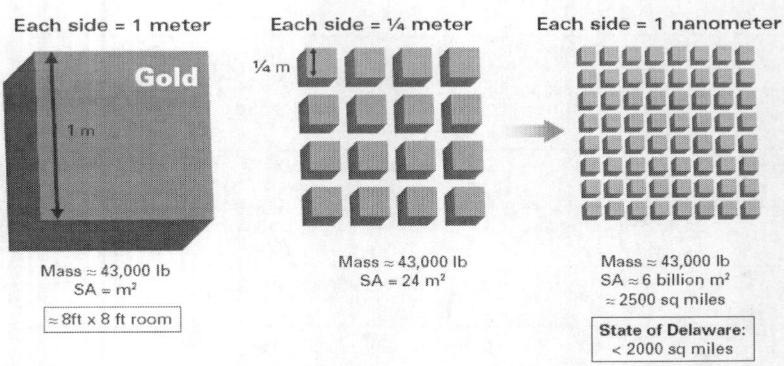

Figure 4. Dividing a solid into nanometer-sized particles exposes a high fraction of the interior atoms to the surroundings making surface area a critical metric for measuring exposure.

Making risk comparisons solely based on weight ignores the importance of surface area and, consequently, may greatly underestimate the health hazards posed by nanoparticles. Referencing 40 year old "nuisance dust" standards, as most current Material Safety Data Sheets for carbon nanotubes do, is essentially saying that normal use of these carbon nanotubes should generate less dust than a sawmill. Almost no other risk conclusion can be drawn from the use of old, mass-based standards. Even if the engineered nanoparticles are delivered suspended in solution, recent studies indicate that workers can be exposed to water droplets containing the particles during standard laboratory practices like sonicating the liquid to break up agglomeration, which is a common occurrence. [19]

3.2.2. Absence of a Permissible Exposure Limit

Currently, there are no accepted occupational exposure limits for any nanoparticles; however, there have been 16 Occupational Exposure Limits offered for consideration by various organizations and companies. [20]

NIOSH is about to complete a Current Intelligence Bulletin on titanium dioxide (TiO_2) that will provide Recommended Exposure Limits (RELs) for two size ranges of TiO_2: one for particles in the micrometer-diameter range and one for particles in the diameter range below 100 nanometers. NIOSH will be providing Recommended Exposure Limits (RELs) for ultrafine (including nanoscale TiO_2) and fine TiO_2. The RELs will be 8-hour TWAs in the range of 0.3 mg/m3 for ultrafine TiO_2 and 2.4 mg/m^3 for fine TiO_2. Nevertheless, NIOSH admits, that questions remain about specifying the sampling and analytical methods for the nanoscale TiO_2, the extent of workplace exposures, and the ability to control exposures at or below the REL." [21]

NIOSH has announced its intention to create one or more Recommended Exposure Limits (RELs) for nanoparticles likely to be commercially available. This may have to be set by broad category of nanoparticles based on physicochemical similarities. There may be, for instance, more than 10,000 combinations of carbon nanotubes possible. [22]

NIOSH, in its draft Current Intelligence Bulletin on carbon nanotubes is recommending an REL of 7 ug/m3 as an 8-hour time weighted average. This is roughly 3 orders of magnitude lower than the levels found in section 2 of many MSDSs for carbon nanotubes now on the market.

Given the tentative nature of governmental action thus far, private companies have begun to set exposure limits for their products. In November 2009, Bayer Material Science (BMS) announced an occupational exposure limit of 0.05 mg/m3 for Baytubes, a multi-wall carbon nanotube product. The company has incorporated this OEL in their Material Safety Data Sheets, which is an enlightened policy compared to the standard practice of referencing Threshold Limit Values for graphite.

Given the lack of benchmarks for sampling, very few private organizations have reported monitoring their workplace for nanoparticles, although those that handle larger volumes of nanomaterials are more likely to do so. [23]

3.2.3. Approaches Used by NIOSH to Count Particles and Measure Surface Area

NIOSH has examined the following innovative approaches in its 2009 guidance document, "*Approaches to Safe Nanotechnology*." [24]

The first is counting particles. Fortunately, the industrial hygiene field has had access to affordable real-time instruments that count particles for a considerable time. The best known use of these instruments is for quantitative fit-testing of respirators with instruments like the TSI Portacount. NIOSH researchers have used handheld **condensation particle counters (CPC)** to count nanoparticles; these instruments use isopropyl alcohol to coat particles so they are large enough to be counted with a laser beam. The CPC report the total number of particles counted per cubic centimeter of air without identifying the chemical makeup. The CPC measures particles in the size range of 10 or 20 nanometers (nm) up to 1,000 nm.

A slight variation on this technique, optical particle counters (OPC), use laser light scattering to provide the total number of particles per liter of air without providing any chemical identification. The OPC measures the total number of particles per liter within 6 specific size cut points: 300 nm; 500nm; 1,000 nm; 3,000 nm; 5,000 nm and 10,000 nm.

NIOSH conducted field studies at 12 sites including research and development laboratories, pilot plants, and manufacturing facilities using a new approach involving a battery of measurements that it called Nanoparticle Emission Assessment Technique (NEAT). The results demonstrated the success of the sampling strategy, using cost-effective portable methods and equipment available to the average industrial hygienist. [25]

A more complex and expensive instrument is the **Particle Surface-Area Analyzer**. Instruments like the TSI Aerotrak™ 9000 Nanoparticle Aerosol Monitor do not measure total active surface area, but indicate the surface area of particles which may be deposited in the lung in units of square micrometers per cubic centimeter, corresponding to either the tracheobronchial or alveolar regions of the lung. The Ecochem DC 2000-CE measures the total particle surface area. These devices are being evaluated by NIOSH for usefulness in conducting initial assessments.

Scanning Mobility Particle Sizers report particle diameter sizes and numbers, which can significantly enhance the capacity to identify releases of engineered nanoparticles, rather than naturally-occurring ultrafines. The SMPS is widely used as a research tool for characterizing airborne nanoparticles, but won't be widely used to measure worker exposures because it is much more expensive and physically larger than other instruments. It also contains an internal radioactive source, which further complicates its use. The cutting edge of sampling may be represented by the **Fast Integrated Mobility Spectrometer (FIMS)**, which has been developed for rapid aerosol size distribution measurements including those aerosols with low particle number concentrations. Results from this instrument compared well with those measured by a scanning mobility particle sizer (SMPS) and total particle concentration measured by the FIMS agreed well with simultaneous measurements by a condensation particle counter (CPC).This device is also able to capture the size distribution of rapidly changing aerosol populations. [26]

3.2.4. Results from Limited Sampling

The results from the NIOSH field survey of 12 facilities reviewed earlier is the most comprehensive look at potential worker exposure published thus far.

An important case study measured very fine particle number and mass concentrations in an engine machining and assembly facility. A condensation particle counter (CPC) and an optical particle counter (OPC) were used to measure particle number concentrations over a broad range. The OPC measurements were used to estimate the respirable mass concentration. The study demonstrated the importance of considering all sources of very fine particles. The authors reported that, "In summer, the very fine particles present in the outdoor air may have substantially contributed to very fine particle number concentration observed inside the plant." In winter the main source of very fine particles was not plant operations but gas-fired heaters. [27]

A NIOSH case study that examined potential emission sources of engineered nanoparticles (ENM) during a variety of operations showed that reactor cleanout was an uncontrolled source of emissions, apparently due to technicians brushing and scraping unwanted buildup from the inside of the reactor. This prompted an effort to minimize potential worker exposure, mainly through the use of PPE as well as consideration of other measures such as local exhaust ventilation systems (LEVs).

Sampling at this site showed that reductions of 96 percent with particle counters and 88 percent with filter-based methods led researchers to report "properly maintained LEV can be highly effective in controlling ENM emissions. This finding, coupled with the current use of PPE, appears to be an acceptable method of reducing the potential for worker exposure." [28]

NIOSH received a request to evaluate a university-based lab that used carbon nanofibers (CNFs) to produce polymer composite materials. NIOSH researchers used various methods to measure exposures during tasks such as transferring CNFs into a lab hood, mixing CNFs with acetone in a vessel and cutting composite with a wet saw. Measurements of total airborne carbon were significantly elevated over background during handling of bulk materials and cutting composite. Surface sampling also suggested CNFs were being transported into adjacent areas, probably on soles of shoes. The authors concluded that the potential for release of engineered nanoparticles does exist during various processes. [29]

Sampling in an automotive grey iron foundry by NIOSH researchers demonstrated the importance of particle concentration mapping because concentrations differed greatly over time and by location, elevating considerably during melting and pouring operations. [30] A separate study in an automotive engine manufacturing plant looked at the relationship among particle number, surface area and respirable mass, and recommended "simultaneous measurements of particle number, active surface area and mass concentrations." [31] If this combination of measurements really is the minimum required to clearly determine worker exposures, then adequate monitoring will not be practicable in many workplaces producing nanoparticles for the foreseeable future, which points out the need for alternative approaches like such as control banding. Fortunately, technology for measuring nanoparticles is constantly improving and advancing, including the development of sensors. (See Section 5.3.5)

3.3. Controlling Exposures

3.3.1. Hierarchy of Controls

Any discussion of training must start with the clear acknowledgment that in the hierarchy of controls, training is an administrative control that is below elimination, substitution and engineering controls in the hierarchy.

The hierarchy has served the industrial hygiene profession well for decades and works well in considering nanoparticles, as indicated in the diagram from NIOSH researchers in Figure 5. NIOSH guidance on applying the hierarchy to nanotechnology can be found in "Approaches to Safe Nanotechnology."

Detailed guidance on applying the hierarchy of controls can be found in the Canadian document, *Best Practices Guide to Synthetic Nanoparticle Risk Management* (IRSST Report 599). The guide recommends that high-risk operations be isolated in separate rooms, ventilated and equipped with independent ventilation systems to avoid the possibility of workstation contamination and worker exposure. A closed circuit process was recommended as the main production method capable of effectively controlling emissions. Carbon black, silica fumes, nanoscaled TiO_2, metals and metal oxides are normally synthesized in closed circuit, according to the Canadian guidance. [32]

An international survey by the International Council on Nanotechnology (ICON) of manufacturing firms and research labs found that the principal means of controlling exposure are:

- 43% laboratory hoods,
- 32% glove boxes,
- 23% vacuum systems,
- 23% white rooms, clean rooms
- 20% closed circuits,
- 15% laminar flow ventilation tables,
- 12% biosafety cabinets and
- 12% glove bag.

Most companies or laboratories use more than one means of emission control so the percent sums to greater than 100. The primary finding of the survey was that "actual reported EHS practices… do not significantly depart from conventional safety practices for handling chemicals." [33]

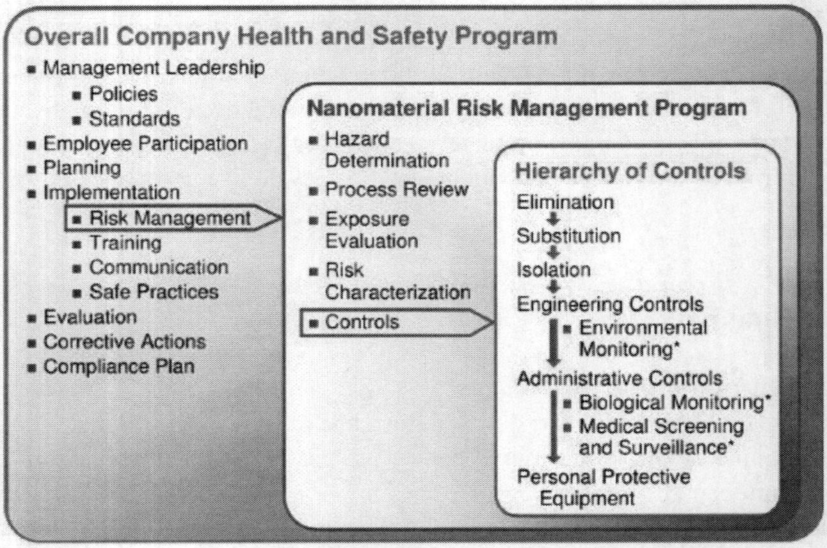

Schulte et al., 2008.

Figure 5. Management system for nanotechnology.

3.3.2. Ventilation

Standard industrial ventilation approaches must be carefully considered because of the buoyancy of nanoparticles. Even when the ventilation is a laboratory fume hood, totally enclosed on three sides, the universal recommendation of a face velocity at the sash of 100 feet per minute will generate too much turbulence inside, possibly releasing particles.

Field studies have shown that handling dry nanoparticles inside laboratory fume hoods can cause a significant release from the hood. Hood design affects the magnitude of release. With traditionally designed fume hoods, the airflow moves horizontally toward the hood, but becomes turbulent in the worker's wake, which can cause nanoparticles to be carried out with the circulating airflow.

> Airborne particle concentrations were measured for three hood designs (constant-flow, constant-velocity, and air-curtain hoods) using manual handling of nanoalumina particles. The hood operator's airborne nanoparticle breathing zone exposure showed high variability for the constant-flow hood while the constant-velocity hood showed some variability, but was usually very low. The performance of the air-curtain hood, a new design with significantly different airflow pattern from traditional hoods, was consistent under all operating conditions and release was barely detected. Fog tests showed more intense turbulent airflow in traditional hoods, but not in the air-curtain hood. [34]

NIOSH, based on field sampling, considers engineering controls that most employers have readily available useful in minimizing nanoparticle emissions. [35]

Manufacturers are now making hoods recommended for use with nanoparticles, but a worldwide survey of laboratories conducting nano-related research found that only 10 percent of researchers reported using nano- enabled hoods, and one in four did not use any type of general laboratory protection. [36]

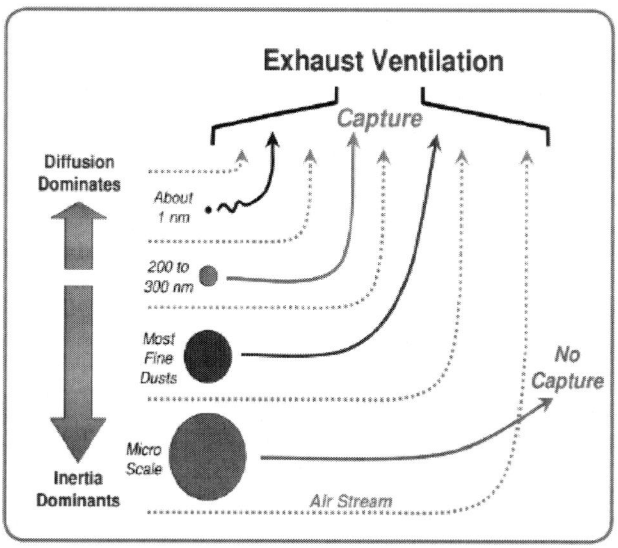

Schulte et al. 2007.

Figure 6. Exhaust ventilation and particle size.

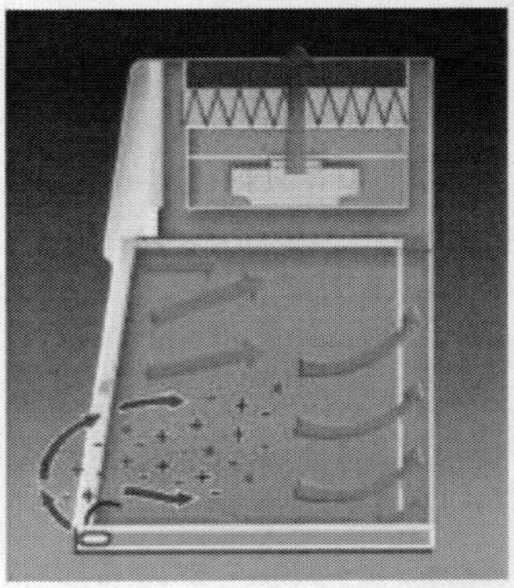

Images courtesy of Labconco.

Xpert Nano Enclosure. Room air is pulled into the enclosure through the front, flows to the baffle and finally passes through a 99.999% ULPA exhaust filter before returning to the laboratory or cleanroom.

3.3.3. HEPA Filtration

Given the extremely small size of nanoparticles there has been understandable concern that they might be able to slip through even the highest efficiency filters. All indications thus far show that this is not the case.

It is important that trainers provide students with a clear understanding of the principles of HEPA filtration. High Efficiency Particulate Air filters are tested and shown to be at least 99.97% efficient against monodispersed aerosols of 0.3 microns in aerodynamic diameter. This does not mean that particles smaller than that size – including all nanoparticles – will pass through the filter, like dust through a screen door. HEPA filter material does not resemble most regular filters, rather forcing the air to follow a "torturous path." The 0.3 micron diameter was chosen because it is the most difficult size to capture, i.e. larger particles will be captured more easily through impaction and smaller ones through electrostatic charges.

NIOSH has concluded that "a well-designed exhaust system with a high-efficiency particulate air (HEPA) filter should effectively remove nanoparticles." [37]

3.3.4. Personal Protective Equipment (PPE)

Despite occupying the bottom rung on the hierarchy of controls, PPE against nanoparticles seems prudent even as other risk management strategies are put in place. Unfortunately, there is little guidance on selecting PPE against nanoparticles. NIOSH researchers have identified that large uncertainties remain, particularly the possibility of a thermal rebound effect for particles as large as 20 nm, as well as the high potential of inward leaks at interfaces. In terms of protective clothing and gloves, the paucity of information available is troubling because the commonly held belief that the skin serves as a barrier to nanoparticles has been thrown into doubt by recent research. [38]

Studies on the filtration performance of N-95 filtering-facepiece respirators have found that the mean penetration levels for 40 nm particles range from 1.4% to 5.2%, indicating that 95 and higher performing respirator filters would be effective at capturing airborne nanoparticles. [39,40,41] A NIOSH approved filtering-facepiece respirator or elastomeric half-face respirator equipped with a 95 or 100 series filter, should provide adequate protection when properly fit-tested on the worker. [42] However, selection of the appropriate respirator type should be based on knowledge of the hazard, the airborne exposure concentration, and whether an exposure limit exists for the engineered nanoparticle. Some trainers associated with the NIEHS WETP program have expressed concerns about the actual protection afforded in the field by disposable filtering-facepiece respirators, compared with more substantial, elastomeric, dual-cartridge respirators. This is a subject for additional research – and meaningful class discussion.

> "It is likely that NIOSH-certified respirators will be useful for protecting workers from nanoparticle inhalation when properly selected and fit-tested as part of a complete respiratory protection program."

One positive sign is that some clothing standards incorporate testing with nanoscale particles and therefore provide some indication of the effectiveness of protective clothing with regard to nanoparticles. [43] One governmental body recommended that outerwear be modified to reduce the production of static electricity, which increases the attraction of nanoparticles. They also recommended the use of disposable protective clothing because cleaning of garments to remove nanoparticles has not been sufficiently evaluated. [44]

The ICON 2006 survey of international firms and labs found that:

- 41 percent of the organizations said they used lab coats (7 percent of which were disposable);
- 26 percent used more protective coveralls (7% of them disposable),
- 11% used shoes reserved for the laboratory,
- 9% have their own laundry service, and
- The most common gloves were nitrile, latex and rubber.

As all trainers know, the efficiency of the respirator can approach zero if the device isn't conscientiously worn. An international survey of nano-related laboratories found that nearly half of the researchers reported not using any type of respiratory protection. [45] Hopefully, the prevalence of respirator use is higher in industry.

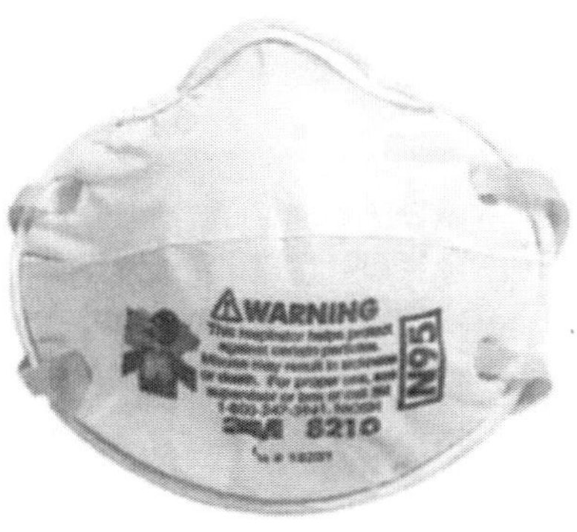

Image courtesy of 3-M Corporation.

N-95 respirator.

3.3.5. Controlling Safety Hazards Like Fire Potential

While most attention has rightly been given to the potential health effects of nanoparticles, there is clearly a need to focus on the safety issues as well. OSHA has begun the process of holding hearings on a combustible dust standard, because more than 130 workers have been killed and more than 780 injured in combustible dust explosions since 1980. [46]

It is common knowledge that explosive dust clouds can be created from most organic materials, many metals and even some non-metallic inorganic materials. The main element affecting the ease of ignition and explosive violence of airborne dust is the particle size and surface area, which are inversely related (i.e. for the same mass, as the particle size decreases, the surface area increases). The violence of the explosion and the ease of ignition generally increase as the particle size decreases. Consequently, many nanoparticle types have the potential to cause explosions, yet data on fire and explosion hazards of nanoparticles is almost nonexistent. [47]

Despite the cutting edge nature of producing nanoparticles and nanomaterials, these workplaces can be susceptible to the same ubiquitous hazards as plants making brooms. Slips, trips, equipment entanglement and other safety hazards will need to be constantly evaluated and controlled.

3.3.6. Hazard Communication for Nanoparticles

In the United States, the Occupational Safety and Health Administration's Hazard Communication Standard (29 CFR 1910.1200) requires that employers inform their workers of the chemical hazards to which they are exposed and how they should protect themselves. In Canada, the Workplace Hazardous Materials Information System (WHMIS) has the same requirement. The 2006 European REACH (Registration, Evaluation, Authorization and Restriction of Chemicals) initiative on chemical hazard communication is more comprehensive and ambitious than the OSHA requirements and requires chemical manufacturers to follow how their products are used by purchasers. THE EU REACH recently announced (Sept 15, 2010) that it is proposing a specific register for nanomaterials. [48]

The most significant development in hazard communication in decades has been the international effort to create a harmonized system. In 2003, the United Nations adopted the Globally Harmonized System of Classification and Labeling of Chemicals (GHS), which includes criteria for classifying health, physical and environmental hazards; GHS also specifies what information should be included on labels of hazardous chemicals and on safety data sheets. OSHA published a proposed rulemaking on September 30, 2009 to align OSHA's Hazard Communication standard (HCS) with the GHS.

The GHS provides consistency in format. Under the GHS, labels would include signal words, pictograms and precautionary statements. Safety data sheets would have a standardized format of 16 sections based on the ANSI Z400.1 consensus standard. It is valuable to ask if it the ANSI standard really is appropriate for nanomaterials. The last section of the GHS (and ANSI) format is called "other information" and is the only section that is not tightly prescribed and could, therefore, contain specific cautionary language about the nano-sized component in the product. One Hazcom expert offered an example of a warning he created: "Established exposure values do not address the small size of particles found in this product and may not provide adequate protection against occupational exposures." [49]

Material Safety Data Sheets (MSDSs) are required for nanoparticles that meet the definitions of hazardous chemicals under OSHA's Hazard Communication standard. According to a survey of firms in Massachusetts, MSDSs from suppliers are the preferred source of risk information for nanotechnology firms. [50] ANSI has recently published a new combined standard that covers MSDSs (ANSI Z400.1) and Precautionary Labeling (ANSI Z129.1).

Unfortunately, industry hasn't done a good job of communicating the hazards of standard industrial chemicals despite having two and a half decades since the promulgation of OSHA's Hazard Communication standard in 1983 to perfect it. An OSHA-funded 1997 study of the peer-reviewed hazard communication literature indicated broad shortcomings with MSDSs, labels and warnings. [51] Three separate studies found that literate workers only comprehended roughly 60 percent of the health and safety information on sample SDSs. [52,53,54]

A recent review of more current literature regarding the accuracy, comprehensibility and use of MSDSs unfortunately did not show improvements over the 1997 review. Accuracy and completeness were found to be relatively poor: the majority of studies showed that the MSDSs did not contain information on all the chemicals present and workers showed low comprehensibility because of overly complex language. [55]

NIOSH appears to maintain the most complete collection of MSDSs for engineered nanoparticles and recently analyzed 60 of them from 33 different manufacturers for technical sufficiency. [56] The researchers only rated 5 percent as "good" while 55 percent were rated as "in need of serious improvement." Over half contained Occupational Exposure Limits (OELs) for the bulk material without providing guidance that the OEL may not be protective for the nanoscale material. Eighty percent "failed to recognize the material as being nano in size or list a particle size distribution showing the nano size range" and a higher percentage "lacked toxicologic data specific to the nanoparticles." Eight percent failed to "suggest any type of engineering controls or mechanical ventilation."

These findings corroborated a similar analysis of a subset of the same MSDSs presented at an international conference sponsored by the EPA in October 2008. [57] The earlier analysis also noted that of those MSDSs that recommended local exhaust ventilation, 25 percent recommended a face velocity greater than 100 feet per minute even though, as noted earlier, NIOSH has specifically warned against operating fume hoods at that rate because the turbulence can release nanoparticles. Additionally, not one of the MSDSs reported that nanoparticles pose a much greater flammability risk despite warnings from authoritative sources that "an increasing range of materials that are capable of producing explosive dust clouds are being produced as nanopowders." [58]

4. Regulatory and Voluntary Approaches Specific to Nanoparticles

Given the many unknowns regarding nanoparticles, a wide range of risk assessment approaches has been suggested for nanotechnology, from recommendations for a total moratorium on any development and use of nanoparticles if and until they are proven to be safe to humans and the environment, to recommendations for relying on existing occupational safety and health laws and regulations. [59]

It is useful to review approaches that are being taken by government as well as new initiatives or paradigms being developed outside a strict regulatory framework.

4.1. Pro-Active Efforts of the Federal Government Compared to Past

Most HAZWOPER trainers cover in their courses, however briefly, the sad and repeated history of occupational diseases killing hundreds or thousands of workers before the federal government acted to eliminate or curtail exposures. Asbestos is the most notorious example, but the list includes acrylonitrile, benzene, acetyl and lead. The latter is painfully illustrative. Austria, France and Belgium phased out lead in household paint in 1909. In this country, the trade association for lead paint agreed to remove it from paint used on children's toys in 1936, but it wasn't phased out completely until 1977. These additional 68 years of exposure from all sources, including leaded gas and house paint, have been credited with causing a population-wide IQ drop. [60] Even with the bans and existing regulations, products containing lead-based paint still surreptitiously enter the country.

The federal government has acknowledged this sorry history and publicly vowed to avoid a similar path with nanotechnology. The results thus far have been encouraging. Under the National Nanotechnology Initiative, all of the key agencies – NIOSH, EPA, OSHA, DOE, DoD, CPSC, FDA – and the White House have been working together in the Nanotechnology Environmental and Health Implications (NEHI) working group to identify the key research that should be conducted to protect workers and the environment.

Government must be constantly on guard, however, for what Princeton historian Ed Tenner called the "tendency of advanced technologies to promote self-deception." According to David Rejeski, Director of the Project on Emerging Nanotechnologies at the Woodrow Wilson International Center for Scholars,"Nanotech is riding the hype wave like a happy surfer at Waikiki," which increases the chance of self-deception because nanotechnology has turned into a "surrogate indicator of U.S. technological leadership in the global economy." [61] He concluded that "Even if some governmental official believed that our existing set of safeguards and statutes would likely fail if applied to nanotechnologies, the probability that they would publicly state such a propostion is infinitesimally small." He also warned that "Dealing with safety issues around nanotechnology at this point in time is a piece of cake compared to what is coming." [62]

4.2. Review of Government Regulatory Actions

4.2.1. Overview

As a set of technologies, materials and devices, "nanotechnology" has been or will be determined to fall within several different regulatory frameworks. Each agency with regulatory authority over some form of nanotechnology has taken concrete steps to understand how best to apply its statutes to this new class of technologies. The actions taken by the key regulatory agencies are summarized in Table 4.

Table 4. Actions on nanotechnology by key federal regulatory agencies

Agency	Primary Statutes	Actions
EPA	Toxic Substances Control Act (TSCA) Federal Insecticide, Fungicide and Rodenticide Act (FIFRA) CERCLA (Superfund) Clean Air Act Resource Conservation and Recovery Act	• Implemented voluntary reporting program (TSCA) • Published white paper (TSCA) • Issued rules specific to nanoparticles (TSCA) • Issued fines for noncompliance (FIFRA) • Funded intramural and extramural research
FDA	Federal Food, Drug and Cosmetic Act	• Created topics page at website • Commenced internal research program • Formed Task Force • Issued white paper • Published monograph on nanoscale sunscreen ingredients
OSHA	Occupational Safety and Health Act	• Created topics page at website • Funded development of worker training materials
CPSC	Consumer Product Safety Act Federal Hazardous Substances Act	• Published white paper • Commenced internal research program

For nanoparticles, the two federal statutes that have the most relevance today are the Toxic Substances Control Act (TSCA) and the Federal Insecticide, Fungicide and Rodenticide Act (FIFRA), both enforced by the Environmental Protection Agency (EPA).

4.2.2. Nanoparticles As Toxic Substances

Nanoparticles meet the definition of chemical substances under TSCA. [63] At issue is whether they are considered new forms of existing chemicals, which would lower the burden on the manufacturer bringing that substance to market, or whether they are "new chemicals", which imposes greater reporting requirements. EPA's initial position that a nanoscale form of a chemical substance is not new solely by virtue of its size has garnered significant criticism. There are indications that EPA may be rethinking this, which could lead to big changes in how the agency regulates nanoparticles. If EPA were to deem nanoparticle forms new chemicals, the nanoparticles could be subject to reporting and testing that are not required for most nanoparticles being sold today.

Meanwhile, EPA is exercising its authority to regulate some nanoparticles using the significant new use rules (SNUR) under Section 5(a)(2) of the Toxic Substances Control Act (TSCA). SNURs have already been issued on certain carbon nanotubes and siloxane nanoparticles, largely based on the nanoscale form and influenced by publicly available toxicity data. The (siloxane) SNURs require manufacturers to notify EPA 90 days prior to sale and describe specific worker protection measures that must be taken when handling the materials. [64] EPA must consider nanoparticles in media other than air; those in liquids will enter the body by different routes and at different rates.

4.2.3. Nanoparticles As Pesticides

The Federal Insecticide, Fungicide, and Rodenticide Act (FIFRA) provides EPA the authority to control the distribution, sale and use of pesticides and pest control devices. Under FIFRA, EPA may determine the risks and benefits of pesticidal products incorporating nanotechnology ("nanopesticides") and may impose restrictions to limit potential risks. Some nanoscale materials such as titanium dioxide and nanoscale silver exhibit potent antimicrobial effects, which have been exploited in a growing number of consumer products. EPA has taken note of several products making antimicrobial claims on the label and leveled fines against the manufacturers for failing to register the products as pesticides or pest control devices as required by FIFRA. Examples of these include a $208,000 fine imposed on ATEN Technology, Inc. for the failure of its subsidiary IOGEAR to register several antimicrobial computer mouse and mouse/keyboard combination products and for making "unverified claims that coatings on keyboard and mouse accessories would eliminate pathogens and kill bacteria," [65] and a fine of nearly $1 million against VF Corporation for failure to register 70 styles of footwear sold under the North Face brand that contain an AgION silver treated foot bed that claim to "inhibit the growth of disease-causing bacteria," "prevent bacterial and fungal growth" and continuously release antimicrobial agents. [66] In each case a nanoscale ingredient is believed to be the active antimicrobial agent. Partly in response to criticism that EPA was not doing enough to scrutinize and regulate nanosilver pesticides, [67] the agency convened a special meeting of its Scientific Advisory Panel (SAP) in November 2009 to get advice on the science behind its regulatory decision making with regard to nanoparticle-based pesticides. As the market for these products continues to grow, the evolution of EPA's thinking on these issues is worth watching.

Their thinking may be prodded by the Government Accountability Office's recent and formal recommendations to the agency to:

1. Complete its plan to issue a Significant New Use rule for nanomaterials.
2. Modify FIFRA pesticide registration guidelines to require applicants to identify nanomaterial ingredients in pesticides.
3. Complete its plan to clarify that nanoscale ingredients in already registered pesticides, as well as in those products for which registration is being sought, are to be reported to EPA and that EPA will consider nanoscale ingredients to be new. [68]

HOW A WASHING MACHINE BECAME A PESTICIDE

In everyday language, a pesticide is something that kills cockroaches, ants, mosquitoes and other nuisance pests. In addition to insects, the word 'pest' in EPA jargon includes another class of 'bugs': microbes. Therefore, any product that claims to kill bacteria, viruses, fungi or other unwanted microbes, is classified as a pesticide or pest control device by the EPA. A product that uses physical or mechanical means to control a pest is a pest control device (e.g., untreated flypaper and UV light disinfection systems) and does not require registration under FIFRA, but if it uses a substance to control pests, then it is a pesticide and must be registered if it makes a pesticidal claim.

In 2005, EPA advised a washing machine manufacturer (presumably Samsung) that its product (presumably the Silver Nano™ Silver Wash) would be classified as a pest control device. The Samsung Silver Wash product claims to kill odor-causing bacteria on fabrics by using electrolysis of a silver electrode to release silver ions into the wash water. Moreover, the product's marketing claims that the silver ions permeate the fabrics, providing anti-bacterial protection for up to one month. [69] Shortly after its decision became public, EPA received letters from waste water treatment facility operators urging it to reconsider its decision and classify the Samsung washing machine as a pesticide because it uses a substance (silver ions) to kill bacteria. [70] The basis for their concern was the inevitable release of silver ion-containing water into sanitary sewer systems which could hamper efforts by plant operators to keep their effluents in compliance with federal limits on silver. Classification of the machine as a pesticide would permit EPA to request data on the potential impact of the machine on silver levels in the waste treatment system and open the door for mandatory restrictions on the sale of the product to avoid further bioaccumulation of silver in the environment. In 2007, EPA revised its ruling "because these items incorporate a substance or substances that accomplish their pesticidal function." [71] And that is how a washing machine became a pesticide.

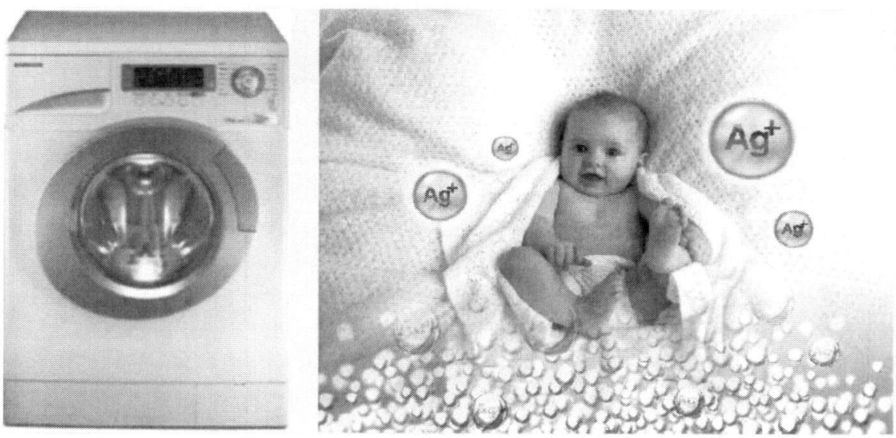

Source: http://www.samsung.com/au/silvernano/site.html.

(L) Samsung's patented Silver Nano™ Silver Wash machine as marketed by Samsung Electronics Australia Pty Ltd. (R) Samsung claims the product "protects sensitive skin and help prevent dermatitis."

4.2.4. Nanoparticles As Workplace Toxicants

Worker exposures have not been directly addressed in regulations, nor does OSHA have any plans to regulate nanoparticles. NIOSH has taken the lead by examining the measurement and control of exposures, but has not yet exercised their powers under the OSHAct to create a recommended standard. When examining the regulatory tools available to OSHA, their 29 CFR 1910.120 HAZWOPER standard must be seen as extremely durable in its applicability over the 20 years of its history - being applied to cleanup of drums of waste in the 90s and the response to terrorist attacks involving anthrax in 2001.

The Hazardous Waste Operation and Emergency Response (HAZWOPER) standard, which went into effect in 1990, protects the safety and health of employees involved in cleanup operations at hazardous waste sites; operations at hazardous waste treatment, storage and disposal facilities; and emergency responses to releases or potential releases of hazardous substances. Although created to protect workers dealing with uncontrolled industrial chemicals at that time, it takes little imagination to see the HAZWOPER connection to nanoparticles. Workers may encounter nanoparticles in the course of handling hazardous waste from a research lab or industrial operation or when cleaning up a large-scale spill in a factory or by the roadside. In addition to these more predictable scenarios for encountering nanoparticles, hazardous waste workers may find themselves purposefully introducing nanoparticles as part of an EPA Superfund cleanup: nanoremediation is projected to be a major part of the overall cleanup strategy on governmental sites over the next 30 years, estimated to be between $87 and $98 billion in scope. The U.S. GAO has recently reviewed the EPA's regulatory options under CERCLA (Superfund) for regulating nanoparticles and noted that the EPA has the statutory authority to designate additional substances as hazardous under CERCLA if their release may present substantial danger to the public health or welfare or to the environment. [72] It may be more a matter of agency commitment and philosophy than regulatory authority.

HAZWOPER trainers may want to visit the EPA's searchable database called Nanotechnology Project Profiles that can be found on its CLU-IN website. [73]

If HAZWOPER coverage involves some ambiguity, OSHA's Hazard Communication standard, 29 CFR 1910.1200, applies completely and importantly. It is hard to envision any population that has a more compelling issue of hazard communication than workers creating nanoparticles or adding them to the broad array of products currently on the market.

Going back to the earliest, contentious debates over the OSHAct, labor and management leaders at least agreed that occupational diseases presented the most serious case for government action. [74] With nanoparticles there are clear safety issues, particularly the risk of fire and explosions from reducing materials to extraordinarily small particle sizes that take exponentially less energy to ignite. [75] But the unknown health risks posed by nanoparticles are an even more stark argument for government involvement.

Properly informing workers about the risks of nanoparticles requires a frank appraisal of the history and current state of industrial chemical regulation. OSHA has regulatory standards, called Permissible Exposure Limits (PELs), for approximately 600 chemical substances; the majority of these PELs are based on consensus standards set at least 40 years ago by volunteer members of the American Conference of Governmental Industrial Hygienists (ACGIH), many of whom were from the industries that produced the chemicals. During the past 40 years, the ACGIH has lowered many of their recommended exposures limits based on continuing research, but OSHA continues to use the levels from 1969, despite an updating of the PELs in 1989 that the courts threw out in 1992.

The age of the OSHA PELs is a minor issue compared to the dearth of information on the overwhelming majority of chemicals in production. There is no definitive count of the number of chemicals in regular use today, but the EPA maintains a list of 83,000 chemicals under the Toxic Substances Control Act. [76] The Chemical Abstract Service had registered 52,122,026 organic and inorganic substances developed by industry as of February 12, 2010. [77]

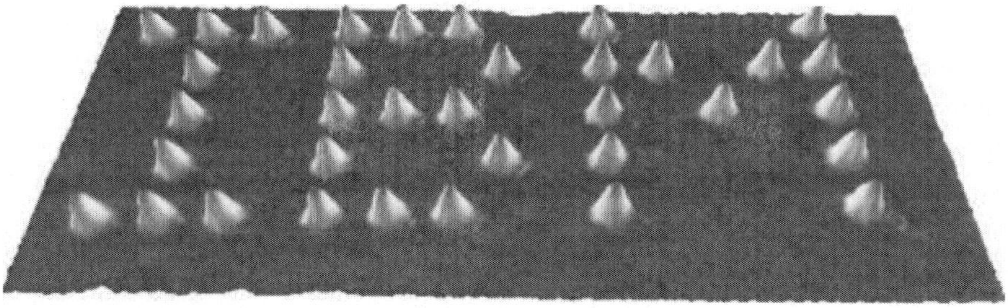

Image courtesy of Wikimedia.

Manipulation at the atomic level by IBM researcher Donald Mark Eigler. In 1989 he spelled IBM with individual xenon atoms using the Scanning Tunneling Electron Microscope.

Given the 118 elements available for combination, a mind-numbing range of between 10^{200} to 10^{900} distinct nanoscale creations has been estimated as plausible; these are truly awe-inspiring numbers for regulators. [78] While federal regulators grapple with these large issues, local and state regulators are dealing with the direct impact of nanomaterials in their jurisdictions.

4.2.5. Regulations at the Local Level

In California, the city of Berkeley amended its hazardous-material reporting requirement in December 2006 to include a notification requirement regarding manufactured nanoparticles. [79] In 2009, the California Department of Toxic Substances Control (DTSC) exercised its authority under the Health and Safety Code, Chapter 699, to request information regarding analytical test methods, fate and transport in the environment, and other relevant information from manufacturers of carbon nanotubes. [80] DTSC has indicated its interest in expanding the data call-ins to other types of nanoparticles. These are the only concrete steps taken by a local or State government but there is active interest in nanoparticles in others, including Massachusetts and Wisconsin, that could result in more actions in the future.

4.3. Voluntary Approaches

There is still a perception among many that guidance and existing regulation are not enough to address the knowledge gaps. In response, established international and intergovernmental bodies and corporations are engaging in their own processes to fill the gaps, often with extensive participation from governments that recognize voluntary processes take less time than regulations. Additionally, grassroots groups and consortia are developing interim strategies for managing risk while governments and other established bodies continue to do their work. Several of these organizations have created guidance documents and consensus standards that trainers can use as resources. One key advantage of voluntary efforts is that guidance can be issued much more rapidly than regulatory rule making. The importance of voluntary guidance is underscored by the international survey of research laboratories referenced previously that found nearly half of the labs had no internal rules on handling nanomaterials and another quarter of the respondents weren't aware of any internal regulations. [81]

4.3.1. Organization for Economic Co-Operation and Development (OECD)

The OECD is an intergovernmental organization in which representatives of 30 industrialized countries in North America, Europe and the Asian and Pacific regions, as well as the European Commission, meet to co-ordinate and harmonize policies and work together to respond to international problems through more than 200 specialized committees and working groups.

The Working Party on Manufactured Nanomaterials was established in 2006 to help member countries address the safety challenges of nanomaterials by bringing together more than 100 experts from governments and other stakeholders. The Working Party is tackling the following important issues:

- Developing a database on human health and environmental safety (EHS) research;
- Establishing EHS research strategies for manufactured nanomaterials;
- Testing the safety of a representative set of nanomaterials; and
- Cooperating on exposure measurement and exposure mitigation. [82]

4.3.2. International Organization for Standardization (ISO)

The International Organization for Standardization (ISO) is a non-governmental organization that develops and publishes voluntary consensus standards via a network of national standards institutes of 161 countries. ISO Technical Committee 229 Nanotechnologies (ISO/ TC 229) was formed to develop consensus standards in nanotechnology and currently has four working groups. Working Group 3 is dedicated to developing standards in the health, safety and environmental aspects of nanotechnologies. ISO/TC 229 recently published a Technical Report, ISO/TR 12885:2008, *Health and safety practices in occupational settings relevant to nanotechnologies*, which focuses on the manufacture and use of engineered nanomaterials.

This report was produced in conjunction with the American National Standards Institute and "provides advice for companies, researchers, workers and other people to prevent adverse health and safety consequences during the production, handling, use and disposal of manufactured nanomaterials." [83] NIOSH personnel were heavily involved in creating this Technical Report, which was based on key NIOSH guidance documents, including Approaches to Safe Nanotechnology.

4.3.3. ASTM International

ASTM International (ASTM) is another international non-governmental organization that develops and publishes voluntary consensus standards. Its Committee E56 on Nanotechnology produced an occupational health standard in 2007 titled, Standard Guide for Handling Unbound Engineered Nanoscale Particles in Occupational Settings. [84] The guide outlines six elements for establishing a program to minimize exposures:

1. Establishing management commitment to the control principle;
2. Identifying and communicating potential hazards;
3. Assessing potential unbound, engineered nanoparticle exposures within the worksite;
4. Identifying and implementing engineering and administrative controls for all relevant operations and activities;

5. Establishing documentation; and
6. Periodically reviewing its adequacy.

ASTM premised exposure control in their Guide on the principle "that, as a cautionary measure, occupational exposures to unbound nanoscale particles should be minimized to levels that are as low as is reasonably practicable." The ALARA principle, As Low As Reasonably Achievable, is the foundation for radiation control and very familiar to trainers who provide radiation worker training.

4.3.4. Nano Risk Framework

One of the most widely recognized voluntary approaches came from the unusual partnership of the DuPont company and the Environmental Defense Fund. In June 2007, they jointly launched the Nano Risk Framework as a comprehensive, practical and flexible system to address the potential risks of nanoscale materials. The Framework has been widely cited as best practice for industry and valuable input for government policy.

The framework has six basic steps (as indicated in the diagram):

1. Describe Material and Its Applications
2. Profile Lifecycles
3. Evaluate Risk
4. Assess Risk Management
5. Decide, Document, and Act
6. Review and Adapt

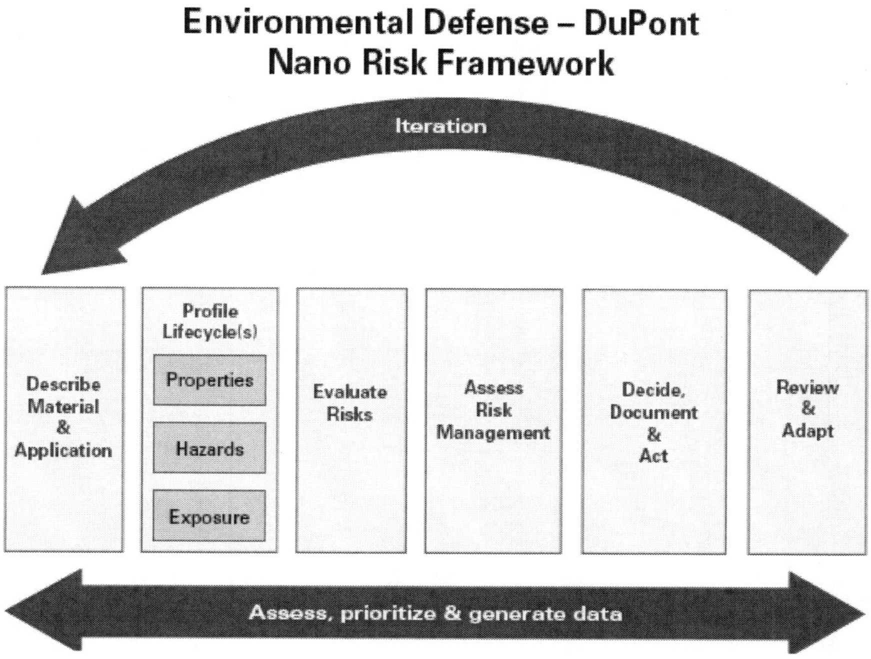

Figure 7. The framework includes a detailed output worksheet that companies can use to document each step of this process.

4.3.5. Control Banding

Given the difficulties, expense and lack of acceptable reference standards for air sampling, as detailed in section 4, another approach is certainly worth pursuing. Control banding has been suggested as a viable option, particularly given its success controlling worker exposures in the absence of complete toxicological and exposure information in the pharmaceutical industry over the last 20 years.

Control banding, unlike air sampling, is a qualitative technique that uses categories or bands of health hazards that are combined with exposure potentials to determine desired levels of control.

A conceptual model was created by Andrew Maynard in 2007 using "impact" and "exposure" indices to combine elements like shape, size and surface area on the nanoparticles with their exposure availability (dustiness and amount in use.) This led to four control strategies:

1. General ventilation,
2. Fume hoods or local exhaust ventilation,
3. Containment, and
4. Seek specialist advice. [85]

A team of experts from Lawrence Livermore National Laboratory elaborated on that design to create a "CB Nanotool" that incorporates a Risk Level (RL) that is a combination of a severity score and a probability score in a standard 4 x 4 risk matrix. The model uses the same four control categories as the Maynard model.

Unlike the Maynard model, however, the CB Nanotool has been validated against the recommendations of independent industrial hygiene experts. The CB Nanotool was demonstrated to have a high level of consistency and tended towards over-control rather than under-control, which is preferable. [86] The tool, which is an Excel spreadsheet, has now been used internationally with good results. It represents an excellent teaching tool, as well. Hazmat instructors should consider having students, preferably working in groups, rate a nanomaterial with which they are familiar using either the electronic CB Nanotool or a paper copy.

The model requires assigning numerical weights to specific severity and probability factors. For instance, surface chemistry must be considered as a severity factor and assigned a score based on whether the surface activity is high (10 points), medium (5) or low (0). Similarly, points are assigned to particle shape, with fibrous forms getting the highest score.

The following severity factors are also scored for the nanoparticles:

- Particle diameter,
- Solubility,
- Carcinogenicity,
- Reproductive toxicity,
- Mutagenicity,
- Dermal toxicity, and
- Asthmagenicity.

Severity \ Probability	Extremely Unlikely (0-25)	Less Likely (26-50)	Likely (51-75)	Probable (76-100)
Very High (76-100)	RL 3	RL 3	RL 4	RL 4
High (51-75)	RL 2	RL 2	RL 3	RL 4
Medium (26-50)	RL 1	RL 1	RL 2	RL 3
Low (0-25)	RL 1	RL 1	RL 1	RL 2

Figure 8. RL 1: General Ventilation, RL 2: Fume hoods or local exhaust ventilation, RL 3: Containment, RL 4: Seek specialist advice.

The probability factors that are weighted and must be reviewed include:

- Estimated amount used during the operation,
- Dustiness of the operation,
- Number of employees with similar exposure,
- Frequency of the operation, and
- Duration of the operation.

Total scores are calculated for severity and probability and the following 4 x 4 matrix is used to determine the Risk Level (RL), which defines the appropriate control strategy. As the Risk Level increases, the control methods similarly increase in protectiveness from general ventilation (RL1), to fume hood or local exhaust ventilation (RL2), to containment (RL3), to seeking the advice of a specialist (RL4). [87]

5. Resources

5.1. Online Nanotechnology Resources for Workers and Trainers

There are many free informational resources on nanotechnology and its potential impacts on human health and the environment. Given the growth of the field of nanotechnology in general and the explosion of information on nano-EHS issues, it is not practical to create an exhaustive list of web-based resources. Instead, the following sites either are themselves essential for all nanotechnology workers and worker trainers or aggregate relevant information from a broad set of sources.

Governmental

- **Environmental Protection Agency (EPA):**
  - National Center For Environmental Research; http://www.epa.gov/ncer/nano/index.html
  - Office of Pollution Prevention and Toxics: http://www.epa.gov/oppt/nano/

- Office of Pesticide Programs: http://www.epa.gov/pesticides/about/intheworks/nanotechnology
- **Lawrence Livermore National Laboratory:** Control banding tool; http://controlbanding.net/Services.html
- **National Institute for Environmental Health Sciences (NIEHS):**
    - Clearinghouse for Worker Safety and Health Training page on nanotechnology; http://tools.niehs.nih.gov/wetp/index.cfm?id=537
    - National Toxicology Program Nanotechnology Safety Initiative; http://ntp.niehs. nih.gov/?objectid=7E6B19D0-BDB5-82F8-FAE7 3011304F542A
- **National Institute for Occupational Safety and Health (NIOSH):** Guidance, field studies, research, Nanoparticle Information Library; http://www.cdc.gov/niosh/topics/ nanotech/
- **National Nanotechnology Initiative (NNI):** US Government's nanotechnology portal; http://nano.gov
- **Oak Ridge Institute for Science and Education (ORISE):** Industrial Hygiene/Occupational Safety Special Interest Group; http://orise.orau.gov/ihos/Nanotechnology/ nanotech_safetyTraining.html
- **Occupational Safety and Health Administration (OSHA):** Standards for occupational practice; http://www.osha.gov/dsg/nanotechnology/ nanotechnology.html

Nongovernmental

- GoodNanoGuide: Information and protocols for safe handling; http://goodnanoguide.org
- International Council on Nanotechnology: Aggregator of nano-EHS news, research, policy reports, industry survey, backgrounders
    - Homepage: http://icon
    - Database to research paper citations: http://icon.rice.edu/virtualjournal.cfm
    - Survey of handling practices in the nanotech workforce: http://tinyurl.com/iconsurvey
- **Project on Emerging Nanotechnologies:** Policy analysis and Consumer Product
- **Inventory:** http://www.nanotechproject.org/
- **Standards**
    - ASTM International Technical Committee E56 on Nanotechnology: http://www.astm.org/COMMIT/COMMITTEE/E56.htm
    - International Organization for Standardization (ISO) Technical Committee 229 Nanotechnologies:
      http://www.iso.org/iso/iso_technical_committee?commid =381983

- **1.4. Educational Repositories**
    - http://www.NanoEd.org (basic to intermediate)
    - http://NanoHub.org (intermediate to advanced)

6. Suggested Training Program

6.1. Limited Literature

Despite the abundance of literature on nanomaterials, only one article on training workers potentially exposed to nanoparticles was identified. The researchers distributed questionnaires and conducted focus groups to assess the training needs of safety and health personnel in nanotechnology industries in Taiwan. [88] They divided training courses into three levels: 1) introductory, which was aimed at providing a basic, awareness level of knowledge for reducing the incidence of occupational illness in nanotechnology industries; 2) advanced, which was aimed at individuals responsible for organizing and conducting occupational illness control programs in nanotechnology industries; and 3) professional, which was aimed at training experts to serve as consultants to reduce incidence of occupational illness in industry. For teaching purposes, these categories correspond roughly to the HAZWOPER categories of first responder awareness, operations level and hazardous materials technician level under 1910.120 (q)(6).

The Taiwanese researchers identified the following training needs based on the level of the individuals:

I. Hazard Recognition		
Introductory Level	**Advanced**	**Professional**
Introduction to nanoparticle health-hazards	Nanoparticle health-hazards of different exposure routes	Health-hazards for nanoparticle inhalatory exposures
	Nanoparticle toxicities and their evaluation techniques	Health-hazards for nanoparticle dermal exposures Nanoparticle toxicities and metabolisms
		Nanoparticle toxicities evaluation techniques
II. Hazard Evaluation		
Introduction to nanoparticle exposure assessments	Inhalation exposure assessment for nanoparticles	Techniques for assessing nanoparticle inhalation exposures
	Dermal exposure assessment	Instrumentation for assessing inhalation exposures
	Biological monitoring for nanoparticles	Principles and techniques used in biological monitoring for nanoparticles
	Sampling strategy for assessing nanoparticle exposure	Sampling strategy for assessing nanoparticle exposure
		Data analysis techniques for assessing nanoparticle exposures
Introduction to the management and control of nanoparticle health hazards	Control of nanoparticle health hazards	Enclosure and isolation techniques for control of nanoparticles

(Continued)

III Hazard Control		
	Management of nanoparticle health hazards	Ventilation techniques for controlling nanoparticle exposures
		Selection of PPE for protection from nanoparticles
		Hazard communication techniques
		Self-auditing techniques for nano-workplaces
		Creation of standard operating procedures for nanoparticle operations
		Medical surveillance concepts for nanoparticle exposures
		Emergency response planning for nanoparticle operations

Suggested Learning Objectives for a Nanomaterials Course for Workers

6.1.1. Module 1: Introduction to Nanotechnology and Nanoparticles

At the end of this module the students will be able to:

- Define nanoparticles and nanomaterials.
- Differentiate among nanoparticles, ultrafines, and engineered nanoparticles.
- Explain the main classes of nanoparticles.
- Describe carbon nanotubes and list some of their valuable properties.
- Explain quantum dots and describe their properties.
- Explain Dendrites and give examples of their properties.
- Analyze the arguments raised about the risks versus the benefits of nanomaterials.
- Describe the main difficulties with characterizing the exposed populations.
- Analyze the importance of considering the life cycle of nanomaterials.

6.1.2. Module 2: Environmental, Health and Safety Impacts of Nanoparticles

At the end of this module the students will be able to:

- Describe the difference between the amount of research on developing nanotechnologies and the amount on the health, safety and environmental impacts of nanotechnologies.
- Describe the routes of entry for nanoparticles into the body.
- Describe several of the health effects caused when nanoparticles enter the body.

6.1.3. Module 3: Application of Traditional Risk Management Approaches to Protect Workers Handling Nanoparticles

At the end of this module the students will be able to:

- List tasks that are most likely to generate worker exposures to nanoparticles.
- Explain why the standard model of industrial hygiene sampling is of questionable value for airborne nanoparticles.
- Explain the importance of the surface area of nanoparticles for biological activity.
- Describe the specific needs for working with nanoparticles under local exhaust contamination.
- Explain the current status of governmental and private efforts to develop occupational exposures limits and analyze the difficulties surrounding those efforts.
- Review the hierarchy of controls and apply it to the management of risk associated with nanomaterials.
- Define High Efficiency Particle Air (HEPA) filtration and describe the current understanding of the effectiveness of HEPA for capturing nanoparticles.
- Discuss the current information available on the protection afforded by NIOSH-certified respirators against nanoparticles, including the use of N-95 filtering facepiece respirators.
- Formulate a message on the effectiveness of protective garments against nanoparticles based on the current findings.
- Describe safety hazards associated with nanomaterial production.
- Characterize the international developments in hazard communication and their impact on informing workers about the risks of nanomaterials.
- List several shortcomings of MSDSs for nanomaterials.

6.1.4. Module 4: Regulatory and Voluntary Approaches Specific to Nanoparticles

At the end of this module the students will be able to:

- Take an informed stance on whether the federal government is being sufficiently proactive managing the risks of nanotechnologies.
- List several regulatory initiatives underway by federal agencies.
- Discuss the difficulties faced by the EPA in regulating nanomaterials that enter the ecosystem.
- Provide an overview of the scope of chemicals in use compared to the number regulated by OSHA and the potential number of chemicals that could be created at the nanoscale.
- Describe several of the international voluntary guidance efforts underway.
- Explain the steps of the Nano Risk Framework for identifying risks of nanoscale materials.
- Explain the principles of control banding and why this approach is receiving serious consideration for assessing the risks of nanomaterials.

6.2. Outline for 8-Hour HAZWOPER Refresher

6.2.1. Purpose

Worker training must be preceded by a needs assessment that allows the training organization to tailor the course as tightly as possible, using workplace examples that are meaningful to the workers being trained. The generic outline provided here should only serve as a template to facilitate adding information needed for protecting the specific worker population. The training should include good adult education techniques, as recommended in the NIEHS Minimum Criteria document (see Section 6.3).

6.2.2. Module 1: Introduction

During this module, the instructor should:

- Review the essential concepts of nanotechnology, particularly the classes of nanomaterials and concepts of relative size.
- Discuss the major areas of application.
- Identify unresolved issues around workforce identification.
- Prompt a discussion about what products the class has observed that contain nanoparticles.
- Ask questions like, "You have been appointed to the worker S&H committee at this carbon nanotube plant, what questions do you ask?" to promote discussions.

6.2.3. Module 2: Environmental, Health and Safety Impacts

During this module, the instructor should:

- Review the major known health effects caused by nanoparticles.
- Describe the limitations to the toxicology information.
- Review the health and safety issues by working through the recognition, evaluation and control of exposures.
- Explain the process of control banding and lead a discussion on why it may be a better approach to controlling exposures to nanoparticles than traditional industrial hygiene.
- Describe the issues around conducting effective hazard communication with limited information.
- Consider disseminating several MSDSs for nanomaterials to facilitate discussions about improving hazard communication to workers.

6.2.4. Module 3: Application of Traditional Risk Management Approaches to Protect Workers Handling Nanoparticles

During this module, the instructor should:

- Facilitate a discussion of possible exposures among the worker populations in class.
- Prompt the class to describe the standard industrial hygiene practice of personal monitoring and why it has serious shortcomings when applied to nanotechnologies.
- Demonstrate the critical importance of surface area in understanding the uniqueness of nanoparticles.
- Compare and contrast the use of local exhaust ventilation for working with standard lab chemicals and nanoparticles.
- Explain the concept of HEPA filtration and engage the class to describe the value of HEPA for nanoparticles.

- Review the current governmental guidance on the protectiveness of N-95 filtering facepiece respirators against nanoparticles and allow the class to debate the use of N-95s.
- Facilitate a discussion on the efforts to develop standards for nanoparticles.
- Challenge the class to think of potential safety hazards associated with nanoparticles and appropriate controls.

6.2.5. Module 4: Regulatory and Voluntary Approaches Specific to Nanoparticles

During this module, the instructor should:

- Describe the efforts of the federal government to identify risk to workers exposed to nanomaterials, and then facilitate a discussion on how those efforts compare with the government's past handling of asbestos, lead, acrylonitrile and other major workplace toxins.
- Review the specific roles of the U.S. agencies responsible for safeguarding workers, the public and the environment from negative effects of nanomaterials.
- Use case studies to illustrate the difficulties faced by EPA in preventing release of nanoparticles into the environment.
- Discuss risks during the life cycle of nanoparticles, from raw materials to inclusion in products to use and to disposal.
- Consider using the internet to show the current number of chemicals assigned a Chemical Asbstract Services number (CAS) and compare that to the number of chemicals in regular use, the number regulated by OSHA and EPA, and to the possible number of nanoparticles that could be generated.
- Proceed systematically through the Nano Risk Framework by Dupont and the Environmental Defense Fund.
- Describe the basic principles of control banding and facilitate a discussion on why this may be the best way to approach controlling nanomaterials. Consider having the students do a group exercise of applying control banding to a chemical with which they are familiar.

Summary of the 2-hour Training Modules

Module	Topic	Description	Adult learning techniques
1	Introduction to nanotech-nology and nanoparticles	An overview of the definition of nanoparticles, classs of nanomaterials with an explanation of engineered versus naturally occurring and a discussion of the benefits, risks and life cycle of nanomaterials.	• Demonstration of commercially available models of nanoparticles; • Group discussions on whether the benefits exceed the risks and who bears the risks;
			• Class exercise of handling actual products (some nano and some not) and trying to determine if they contain nanoparticles; • Class discussion of operations in student workplaces where nanoparticles or nanomaterials are handled.

(Continued)

2	Environmental, health and safety impacts of nanoparticles	A facilitated discussion of routes of entry, known health effects, the value of control banding and areas where research is still needed.	• Group exercises analyzing historical occupational health problems in the workplace and their applicability to nanoparticles; • Demonstration using the Livermore Control Banding Nanotool
3	Application of traditional risk management approaches to protect workers handling nanoparticles	Discussion of limitations of standard industrial hygiene approach to airborne measurement, status of development of occupational exposure limits, current understanding of ventilation effectiveness handling nanoparticles and respiratory protection recommendations by NIOSH. Review of hazcom issues.	• Facilitated discussion of the hierarchy of controls and its applicability to nanoparticles; • Group exercise reviewing actual MSDSs for nanomaterials; • Hands-on exercises with N-95 filtering facepiece respirators; • Hands-on exercises with industrial hygiene sampling equipment.
4	Regulatory and voluntary approaches	Review of regulatory efforts across the federal government and internationally to protect workers from exposure to nanomaterials.	• Class internet exercise putting number of chemicals in perspective by checking current CAS numbers and potential number of nanoparticles versus EPA and OSHA regulated chemicals; • Group exercise applying Dupont/EDF NanoRisk Framework; • Group exercise applying control banding to a specific chemical familiar to the class; • Class discussion on the steps OSHA should be taking; • Class discussion on the value of the Precautionary Priniciple.

6.3. Value of NIEHS Minimum Criteria in Structuring Nanoparticles Training for Workers

There are many good sources available for creating effective training materials for workers. Arguably the guidance with the most substantial results to corroborate its value is the "Minimum Criteria" guidance of the National Institute of Environmental Health Sciences' Worker Education and Training Program. This guidance, which was updated in 2006, has provided the underlying principles for the creation, delivery and evaluation of training for over two million workers since the beginning of the program in 1987.

The initial quality control for the program was developed through a participatory national technical workshop in 1990 and issued by the Program in 1991. This original "Minimum Criteria" was updated in 1994 as the "Interpretive Guidance" to the "Minimum Criteria." The guidance has served as the quality control basis for the WETP training grants program to the present time. It was also adopted by OSHA as a non-mandatory appendix to the HAZWOPER standard.

The following Minimum Criteria recommendations should be applied as much as practicable to any training program created to deal with nanoparticles:

- Provide peer-to-peer training with hands-on activities whenever possible.
- Fill at least one-third of the training program hours with hands-on training.
- Avoid making computer-based training methods the sole form of training, although they can greatly augment the effectiveness and reduce the cost of hazardous waste worker training.
- Make sure proven adult-learning techniques are the core of all worker training.
- Precede all worker safety and health training with a needs analysis to ensure the appropriate knowledge, skills and attitudes are being transmitted.
- Follow all training with a proper evaluation to document that the knowledge, skills or attitudes were acceptably transmitted and that the worker possesses the necessary abilities to perform the tasks. [89]

REFERENCES

[1] Fairbrothers, A. (2008, Oct.) Managing the risks of nanomaterials. International Perspectives on Environmental Nanotechnology Applications and Implications Conference Proceedings, Volume 2 – U.S. EPA Region 5. Chicago, Illinois. Retrieved from: http://www.epa. gov/osp/hstl/Nano2008v2.pdf.

[2] Department of Health and Human Services Centers for Disease Control and Prevention, National Institute for Occupational Safety and Health. (2009, February). *Interim guidance for medical screening and hazardous surveillance for workers potentially exposed to engineered nanoparticles.* (DHHS NIOSH Publication No. 02-2650). Retrieved from http://www. cdc.gov/niosh/docs/2009-116/pdfs/2009- 116.pdf.

[3] U.S. Government Accountability Office. (2010, May). Report to the Chairman, Committee on Environment and Public Works, U.S. Senate. *Nanotechnology: Nanomaterials are widely used in commerce, but EPA faces challenges in regulating risk* (GAO-10-549), 26.

[4] Schmid, K., Danuser B. & Riediker. (2010, April). Nanoparticle usage and protection measures in the manufacturing industry: A representative survey. *Journal of Occupational and Environmental Hygiene*, *7*, 224-232.

[5] Plunkett, J. W. (2009). Nanotechnology & MEMS Industry Almanac 2009 [Electronic version], Plunkett Research Ltd. Retrieved December 2010, from http:// www.plunkettresearch.com/Industries/NanotechnologyMEMs/tabid/198/Default. aspx

[6] Karn B., Kuiken T. & Otto M. (2009). Nanotechnology and *in situ* remediation: A review of the benefits and potential risks. *Environmental Health Perspectives*, *117*(12), 1824. doi:10.1289/ehp.0900793.

[7] Ibid. Karn B., Kuiken T. & Otto M. (2009). 1825.

[8] Ibid. Karn B., Kuiken T. & Otto M. (2009). 1825 1830.

[9] PARS Environmental Inc (2004). NanoFe™: *An innovative remediation technology for soils and groundwater.* Retrieved December 2010, from http://www.parsenviro. com/nanofeaw-1.htm.

[10] Royal Commission on Environmental Pollution. (2008). *Novel materials in the environment: The case of nanotechnology*. London: Royal Commission on Environmental Pollution. Retrieved December 2010 from http://www. official-documents.gov.uk/doc ument/cm74/7468/7468.pdf.

[11] International Council on Nanotechnology. (2008). *Towards predicting nanobiointeractions: An international assessment of nanotechnology environment, health and safety research needs*. Retrieved May 10, 2010 from http://bit.ly/dhphR2.

[12] Ostrowski, A. D., Martin, T., et al. (2009). Nanotoxicology: characterizing the scientific literature, 2000-2007. *Journal of Nanoparticle Research*, *11*(2), 251-257.

[13] Takagi, A., Hirose, A., et al. (2008). Induction of mesothelioma in p53+/- mouse by intraperitoneal application of multi-wall carbon nanotube. *Journal of Toxicological Sciences*, *33*(1), 105-116.

[14] 14 Poland, C. A., Duffin, R., et al. (2008). Carbon nanotubes introduced into the abdominal cavity of mice show asbestoslike pathogenicity in a pilot study. *Nature Nanotechnology*, *3*(7), 423-428.

[15] NIOSH Current Intelligence Bulletin. (2010). Draft report on occupational exposure to carbon nanotubes and nanofibers. http://www.cdc.gov/niosh/ docket/review/ docket161A/pdfs/ carbonNanotubeCIB_PublicReviewOfDraft. pdf.

[16] Ibid., International Council on Nanotechnology. (2008).

[17] Schulte, p., Geraci, C., Zumwalde, R., Hoover, M. & Kuempel, E. (2008). Occupational risk management of engineered nanoparticles. *Journal of Occupational and Environmental Hygiene*, *5*(4), 240.

[18] British Standards Institute. (2007). *Nanotechnologies part 2: Guide to the safe handling and disposal of manufactured nanoparticles*. (PD 6699-2).

[19] Johnson, D. R., Methner, M. M., Kennedy, A. J. & Steevens, J. A. (2010, January). Potential for occupational exposure to engineered carbon-based nanomaterials in environmental laboratory studies. *Environmental Health Perspectives*, *118*(1).

[20] Schulte, P. A., Murashov, V., Zumwalde, R., Kuempel, E. D. & Geraci, C. L. (2010). Occupational exposure limits for nanomaterials: state of the art. *J Nanopart Res*, *12*, 1971-1987.

[21] Department of Health and Human Services Centers for Disease Control and Prevention, National Institute for Occupational Safety and Health (2009, Nov.) *Strategic plan for NIOSH nanotechnology research and guidance: Filling the gaps*. (DHHS NIOSH Publication No. 2010–105). Retrieved December 2010, from http://www. cdc.gov/niosh/docs/2010-105/

[22] Ibid. National Institute for Occupational Safety and Health (2009) v.

[23] Ibid. National Institute for Occupational Safety and Health (2009) 35.

[24] International Council on Nanotechnology. (2006, Nov). *A review of current practices in the nanotechnology industry – Phase two report: Survey of current practices in the nanotechnology workplace*. (University of California, Santa Barbara). Retrieved March 1, 2010 from http://cohesion.rice. edu/CentersAndInst/ICON/emplibrary/ICON NanotechSurveyFullReduced.pdf

[25] Department of Health and Human Services Centers for Disease Control and Prevention, National Institute for Occupational Safety and Health. (2009). *Approaches to safe nanotechnology:An information exchange with NIOSH*. Retrieved December 2010 from http:// www.cdc.gov/niosh/topics/nanotech/ safenano/

[26] Methner, M., Hodson, L., Dames, A. & Geraci, C. (2010). Nanoparticle emission assessment technique (NEAT) for the identification and measurement of potential inhalation exposure to engineered nanoparticles—part B: Results from 12 field studies. *Journal of Occupational and Environmental Hygiene*, *7*(3), 163 – 176.

[27] Olfert, J. S., Kulkarni, P. & Wang, J. (2008). Measuring aerosol size distributions with the fast integrated mobility spectrometer. *Journal of Aerosol Science*, *39*, 940-956. Retrieved from http://www.ecd.bnl.gov/ pubs/BNL-81288-2008-JA.pdf

[28] Heitbrink, W. A., Evans, D. E., Peters, T. M. & Slavin, T. J. (2007). Characterization and mapping of very fine particles in an engine machining and assembly facility. *Journal of Occupational and Environmental Hygiene*, *4*(5), 341 – 351.

[29] Old, L. & Methner, M. M. (2008). Engineering case reports. *Journal of Occupational and Environmental Hygiene*, *5*(6), D63 - D69.

[30] Mazzuckelli, L. F., Methner, M. M., Birch, M. E., Evans, D. E., Ku, Bon-Ki, Crouch, K. & Hoover, M. D. (2007). Case study. *Journal of Occupational and Environmental Hygiene*, *4*(12), D125 - D130.

[31] Heitbrink, W. A., Evans, D. E., Ku, Bon Ki, Maynard, A. D., Slavin, T. J. & Peters, T. M. (2009). Relationships among particle number, surface area, and respirable mass concentrations in automotive engine manufacturing. *Journal of Occupational and Environmental Hygiene*, *6*(1), 19 -31.

[32] Evans, D. E., W. A. Heitbrink, T. J. Slavin, & Peters, T. M. (2008). Ultrafine and respirable particles in an automotive grey iron foundry. *Annals of Occupational Hygiene*, *52*(1), 9–21.

[33] Ostiguy, L., Roberge, B., Ménard, L. & Endo, C. (2009, Jan.) Best practices guide to synthetic nanoparticle risk management. Institut de recherche Robert-Sauvé en santé et en sécurité du travail (IRSST). Retrieved March 1, 2010 from http:// www.irsst.qc.ca/files/documents/ PubIRSST/R-599.pdf

[34] Gerritzen, G., et al. (2006). Review of safety practices in the nanotechnology industry: aurvey of current practices in the nanotechnology workplace. *International Council on Nanotechnology*. http://tinyurl. com/iconsurvey

[35] Tsai, S. J., Huang, R. F. & Ellenbecker, M. J. (2010). Airborne nanoparticle exposures while using constant-flow, constant-velocity, and air-curtain-isolated fume hoods. *Annals of Occupational Hygiene 2010*, *54*(1), 78-87.

[36] Ibid. Methner, M., Hodson, L., Barnes, A., & Geraci, C. (2010)

[37] Balas, F., Arruebo, M., Urrutia, J. & Santamaria, J. (2010). Reported nanosafety practices in research laboratories worldwide. *Nature Nanotechnology*, *1*.

[38] National Nanotechnology Initiative. (2007, March 6). *Precautionary measures for employers and workers handling engineered nanoparticles: NIOSH recommendations*. Retrieved February 10, 2010, from http://www.nano.gov/html/ society

[39] Dolez, P. I., Bodila, N., Lara, J. & Truchon, G. (2010). Personal protective equipment against nanoparticles. *International Journal of Nanotechnology.*, *7*(1), 99-11.

[40] Balazy, A., Toivola, M., Reponen, T., Podgorski, A., Zimmer, A. & Grinshpun, S. A. (2006). Manikin-based performance evaluation of N95 filtering-facepiece respirators challenged with nanoparticles. *Ann Occup Hyg*, *50*, 259-269.

[41] Rengasamy, S., Verbofsky, R., King, W. P. & Shaffer, R. E. (2007). Nanoparticle penetration through NIOSH-approved N95 filtering-facepiece respirators. *Journal International Society for Respiratory Protection*, *24*, 49-59.

[42] Rengasamy, S., King, W. P., Eimer, B. & Shaffer, R. E. (2008). Filtration performance of NIOSH-approved N95 and P100 filtering- facepiece respirators against 4-30 nanometer size nanoparticles. *J Occup Environ Hyg*, *5*(9), 556-564.

[43] Shaffer, R. E. & Rengasamy, S. (2009). Respiratory protection against airborne nanoparticles: a review. *J Nanopart Res*, DOI.10.1007/s11051-009-9649.3.

[44] Ibid. Methner, M., Hodson, L., Barnes, A., & Geraci, C. (2010)

[45] Ibid. Ostiguy, L., Roberge, B., Ménard, L. & Endo, C. (2009, January).

[46] Ibid. Balas, F., Arruebo, M., Urrutia, J, & Santamaria, J. (2010).

[47] U.S. Department of Labor, OSHA Office of Communications. (2010, Jan. 27). *OSHA schedules informal stakeholder meetings on combustible dust*. Retrieved February 8, 2010, from http://www.osha.gov/pls/ oshaweb/owadisp.show_document?p_ table= NEWS_RELEASES&p_id=17112

[48] Ibid. British Standard Institute (2007).

[49] REACH register to ensure traceability of nanomaterials. (2010, 15 September) http://www.euractiv.com/en/foodregister-ensure-traceability-nanomaterials-news-497781

[50] Levine, Dan. (2006, Sep 15). Personal communication.

[51] Lindberg, J. E. & Quinn, M. M. (2007, Dec). A survey of environmental, health and safety risk management information needs and practices among nanotechnology firms in the Massachusetts region. *PEN Research Brief No 1*. Retrieved December 2010, from http://www.nanotechproject. org/publications/archive/a_survey_ environmental_health_ safety/

[52] Sattler, B., Lippy, B. & Jordan, T. (1995). Hazard communication: A review of the science underpinning the art of communication for health and safety. Retrieved December 2010, from http://www.osha.gov/SLTC/ hazardcommunications/ otherresources. html

[53] Kolp, P., Sattler, B., Blayney, M. & Sherwood, T. (1993). Comprehensibility of material safety data sheets. *American Journal of Industrial Medicine*, (*23*), 139.

[54] Phillips, C., Wallace, B., Hamilton, C., Pursley, R., Petty, G. & Bayne, C. (1999). Efficacy of MSDSs and worker acceptability. *Journal of Safety Research*, (*30*), 113-122.

[55] Printing Industries of America. (1990). *Comments on the OSHA Hazard Communications Standard*. Docket H-022G.

[56] Nicol, A. M., Hurrell, A. C., Wahyuni, D., McDowall, W. & Chu, W. (2008, July) Accuracy, comprehensibility, and use of material safety data sheets: A review. *American Journal of Industrial Medicine*, *11*(51), 861-876.

[57] Crawford, C. & Hodson, L. (2009, June). Guidance for preparing good MSDSs for engineered nanoparticles. Poster session presented at the annual meeting of the American Industrial Hygiene Association, Cincinnati, OH.

[58] Lippy, B. E. (2008, Oct.) MSDSs fail to communicate the hazards of nanotechnologies to workers. Paper presented at the International Environmental Nanotechnology Conference, Chicago, Illinois.

[59] Pritchard, D. K. (2004). *Literature Review – Explosion Hazards Associated with Nanopowders*. (HSL/2004/12). Health and Safety Lab. Health and Safety Executive. Harpur Hill, Buxton, UK.

[60] Murashov V. & Howard J. (2009, Aug.) Essential features for proactive risk management. Commentary. *Nature Nanotechnology, 4*, 469.
[61] Markowitz, G. & Rosner, D. (2002). *Deceit and denial: The deadly politics of industrial pollution*. University of California Press: Berkley and Los Angeles, CA.
[62] Rejeski, D. (2006, April 27). Nanotech safety 101 or how to avoid the next little accident. Presentation given at Harvard University's Workshop on Disaster Prevention.
[63] Ibid. Rejeski, D. (2006, April 27).
[64] U.S. Environmental Protection Agency. (2008). *Concept paper for the nanoscale materials stewardship program under TSCA*. Retrieved May 11, 2010, from http://www. regulations
[65] Environmental Protection Agency (2008). *Significant new use rules on certain chemical substances. Federal Register, 73*(215), 65743-65766.
[66] U.S. Environmental Protection Agency. (2008). *U.S. EPA fines southern California technology company $208,000 for 'nano coating' pesticide claims on computer peripherals*. Retrieved December 23, 2009, from http://yosemite.epa. gov/opa/ admpress.nsf/2dd7f669225439b78525735900400c31/16a190492f2f25d585257403005c 2851!OpenDocument.
[67] U.S. Environmental Protection Agency. (2009). *The North Face' clothing parent company facing nearly $1M in federal fines 41 training workers on risks of nanotechnology following unsubstantiated product claims*. Retrieved December 23, 2009, from http://yosemite.epa.gov/opa/ admpress.nsf/2dd7f669225439b785 2 5735900400c31/bcbd9b468b9aaf67852576390055de2f!OpenDocument.
[68] International Center for Technology Assessment. (2008). Petition for Rulemaking Requesting EPA Regulate Nano-silver Products as Pesticides. Retrieved May 30, 2008.
[69] U.S. GAO. (2010). 50.
[70] Samsung. *Silver Nano Health System*. Retrieved December 2010, from http:// www. samsung.com/au/silvernano/site. html.
[71] Weir, C. (2006). *Request to require registration of samsung silver wash clothing washing machine, and silver ion pesticide products*. Retrieved January, 2008, from http://www.tritac.org/ documents/letters/2006_01_27_EPA.
[72] U.S. Environmental Protection Agency (2007). *Pesticide registration; Clarification for ion-generating equipment. Federal Register, 72*(183), 54039-54041.
[73] U.S. GAO. (2010). 41.
[74] U.S. EPA. *CLU-IN technology innovation program*. Retrieved December 2010, from http://www.clu-in.org/products/nano/search.cfm
[75] Mendeloff, J. (1980). *Regulating safety: An economic and political analysis of occupational safety and health policy*. Cambridge, MA: The MIT Press, 13.
[76] Pritchard, D. K. (2004, March). Literature review: explosion hazards associated with nanopowders. *Heath and Safety Laboratory*. HSL/2004/12.
[77] U.S. EPA. (2010). *New chemical program under TSCA*. Retrieved February 7, 2010, from http://www.epa.gov/opptintr/newchems/pubs/invntory.htm.
[78] Chemical Abstract Service. (2010). *Registry Number and Substance Count*. Retrieved February 2, 2010, from http://www.cas. org/cgi-bin/cas/regreport.pl.
[79] Frey, T. (October, 2008). *Nanotech and the precautionary principle*. Retrieved December 2010, from http://www. futuristspeaker.com/2008/10/nanotech- andthe-precautionary-principal/

[80] City of Berkeley (2009). *Manufactured Nanoscale Materials Health & Safety Disclosure*. Toxics Management Division, Planning and Development Department.

[81] California Department of Toxic Substances Control. (2010). Nanotechnology call-in. Retrieved February 2, 2010, from, http://www.dtsc.ca.gov/ TechnologyDevelopment/ Nanotechnology/ index.cfm.

[82] Ibid. Balas, F., Arruebo, M., Urrutia, J. & Santamaria, J. (2010).

[83] Organization for Economic Co-Operation and Development (2009, June). *Series on safety of manufactured nanomaterials, number 10*. ENV/JM/MONO(2009)15.

[84] International Organization for Standardization. (2008). Nanotechnologies -- Health and safety practices in occupational settings relevant to nanotechnologies. http:// www.iso.org/iso/ catalogue_detail?csnumber=52093

[85] *ASTM International. (2007). Standard guide for handling unbound engineered nanoscale particles in occupational settings.* http://www.astm.org/Standards/ E2535.htm.

[86] Maynard, A. (2007). *Nanotechnology: the next big thing, or much ado about nothing? Annals Occupational Hygiene*, *51*(1),1-12.

[87] Zalk, D. M., Paik S. Y. & Swuste, P. (2009, June). Evaluating the control banding nanotool: A qualitative risk assessment method for controlling nanoparticle exposures. *Journal of Nanoparticle Research*, *11*(7), 1685-1704.

[88] Zalk, D. & Paik, S. (2010, March). Control banding and nanotechnology. *Synergist*, *21*(3), 28.

[89] Hsieh, F. M., Tsai P. J., Chen, W. Y. & Chang, C. P. (2009). Developing on-the-job Training Program for the occupational safety and health personnel in nanotechnology industries. *Proceedings of the 2009 IEEE IEEM.*

[90] National Institute of Environmental Health Sciences. *Worker Education and Training Program.* (2006, Jan.) Minimum Health and Safety Training Criteria.

INDEX

A

B

C

D

E

F

J

K

L

M

N

O

P

T

U

V

W

X

Y

Z